Understand Electronics

Understand Electronics

Dr Malcolm Plant

For UK order enquiries: please contact Bookpoint Ltd,
130 Milton Park, Abingdon, Oxon, OX14 4SB.
Telephone: +44 (0)1235 827720. Fax: +44 (0)1235 400454.
Lines are open 09.00–17.00, Monday to Saturday, with a 24-hour
message answering service. Details about our titles and how to
order are available at www.teachyourself.com

For USA order enquiries: please contact McGraw-Hill Customer
Services, PO Box 545, Blacklick, OH 43004-0545, USA.
Telephone: 1-800-722-4726. Fax: 1-614-755-5645.

For Canada order enquiries: please contact McGraw-Hill Ryerson
Ltd, 300 Water St, Whitby, Ontario L1N 9B6, Canada.
Telephone: 905 430 5000. Fax: 905 430 5020.

Long renowned as the authoritative source for self-guided learning –
with more than 50 million copies sold worldwide – the Teach Yourself
series includes over 500 titles in the fields of languages, crafts, hobbies,
business, computing and education.

British Library Cataloguing in Publication Data: a catalogue record
for this title is available from the British Library.

Library of Congress Catalog Card Number: on file.

First published in UK 1988 by Hodder Education,
338 Euston Road, London, NW1 3BH.

First published in US 1988 by the McGraw-Hill Companies, Inc.

This edition published 2010.

Previously published as *Teach Yourself Electronics.*

The Teach Yourself name is a registered trade mark of
Hodder Headline.

Typeset by MPS Limited, A Macmillan Company.

Printed in Great Britain for Hodder Education, a division of
Hodder Headline, an Hachette UK Company, 338 Euston Road,
London NW1 3BH, by CPI Group (UK) Ltd, Croydon, CR0 4YY.

The publisher has used its best endeavours to ensure that the URLs
for external websites referred to in this book are correct and active
at the time of going to press. However, the publisher and the author
have no responsibility for the websites and can make no guarantee that
a site will remain live or that the content will remain relevant, decent
or appropriate.

Hachette UK's policy is to use papers that are natural, renewable
and recyclable products and made from wood grown in sustainable
forests. The logging and manufacturing processes are expected to
conform to the environmental regulations of the country of origin.

Impression number   10 9 8 7 6 5 4
Year   2014 2013 2012

Acknowledgements

I have always enjoyed teaching for there is nothing more rewarding than helping students learn complex or confusing concepts. Of course, when you are in a classroom or lecture room you can always use different ways of explaining things until your students have 'got it'. However, this is much more difficult when you are writing a text since you have only one chance of explaining things. However, I have drawn on my experience of working alongside students and colleagues who have provided the feedback that has helped me organise and present electronic concepts as described in this book. In particular I want to thank my colleague Andy Cooper who is an enthusiastic lecturer having a particular interest in the field of computer control – see Section 13.9 and Taking it Further. I would also like to thank Mike Kenward for allowing me to include in Chapter 16 some of the 19 projects that I prepared for publication in his journal *Everyday Practical Electronics* during 2008/9. This journal is a treasure house of project ideas, sources of components and equipment, and discussion forums, and it is essential reading for both the beginner and the experienced constructor. My thanks also go to Geoff Hallam, some time has passed since we worked together preparing and marketing materials for students, teachers and lecturers but his talent for circuit design and assembly that transformed ideas on paper into practical reality was remarkable. Not least, I want to thank the Teach Yourself team at Hodder Education for their enthusiasm and creative insight, especially Harry Scoble, Development Editor, Hodder Education and Helen Rogers, Editorial Assistant. Finally, I want to thank Brenda who continues to support my various projects, none more satisfying than the happy enterprise of revising this book.

The publisher would also like to thank Everyday Practical Electronics for allowing the use of the following projects, from their monthly magazine, in this book:

Project 1: Circuit Tester   Oct 2009 Vol 37 No 10
Project 2: Dark Switch   April 2009 Vol 38 No 4
Project 3: Games Timer   Dec 2008 Vol 38 No 12
Project 4: Frost Alert   Feb 2009 Vol 38 No 2
Project 5: Rain Check   Oct 2008 Vol 37 No 10
Project 6: Simple Die   Feb 2009 Vol 38 No 2

Contents

Meet the author

Welcome to *Understand Electronics*!

> ***When you know a thing, to hold that you know it; and when you do not know a thing, to allow that you do not know it – this is knowledge.***
>
> Confucius (551–479 BC), *The Confucian Analects*

Electronics is changing the way we live, work and communicate. Digital radios, keyboards and TVs, iPods, digital cameras and GPS (global positioning system) devices are predominantly electronic and they make everyday life more enjoyable, creative and exciting. Indeed, developments in electronics continue to result in new products and services, and it is an essential enabler for innovation and stimulating competitiveness in the global marketplace. This book describes some of these developments, for example in the fields of instrumentation, communications and control, and it will involve you in learning how basic concepts are applied in the design and use of electronic circuits and systems. Whatever your reasons for studying electronics, it is as well to consider that learning about and using electronics is essentially a creative activity and is therefore dependent on human needs and market forces. Electronics can be used for human good as well as in ways that are immoral and destructive, just like any other technology. With this in mind, electronics enthusiasts should resist any temptation to design, make and use electronics devices that harm living creatures and their environments.

In *Understand Electronics* I emphasize how the electrical properties of individual components such as thermistors and transistors help to explain how circuits work. In order to help you, these circuits are considered as a set of building blocks such as switches, rectifiers, timers, amplifiers, oscillators, logic gates and counters. Each chapter begins by setting out learning outcomes that you can expect to achieve by the end of the chapter. Furthermore, to help you grasp the principles and practical applications of electronics each chapter ends with ten revision questions covering the most important things you have learned from reading that chapter. Importantly, in order to endorse the point made above that electronics is essentially about the creative development of things we use, the final chapter describes seven projects that can be assembled using a 'breadboard'. Sufficient guidance is given to ensure that the enterprising reader can acquire the skills to build a working circuit without too much trouble and expense.

Chapter 1 describes some of the uses of electronics, for example, in the home, in medicine, in robotics and in space research, and briefly traces the more significant milestones in the history of electronics. The following four chapters develop a systematic understanding of electronics, starting with ideas about current flow in simple circuits and the electrical properties of conductors and semiconductors that are the basic materials used for making diodes, transistors and integrated circuits. Chapter 5 draws attention to the relationship between current, voltage and resistance, and the use of resistors in controlling the distribution of voltage in simple circuits, which is an essential part of understanding how circuits work.

The purpose of Chapters 6 to 11 is to explain how these basic concepts determine the function of a number of important circuit building blocks. Thus, Chapter 6 shows that electronic timers, used in all kinds of ways from washing machines to cameras, depend on the way capacitors are able to store electrical charge. The function of semiconductor diodes as rectifiers in power supplies is described in Chapter 7. Transistors, the building blocks of audio amplifiers and integrated circuits, are examined in Chapter 8. The digital function of logic gates as building blocks of electronic decision-making circuits is the focus for Chapter 9. Again, the digital function of flip-flops and their applications in electronic counters is the focus of Chapter 10. How light-emitting diodes and liquid crystal devices are used to produce electronic displays is discussed in Chapter 11. The fascinating field of microelectronics is the focus of Chapter 12.

It includes the use of flip-flops in computer memories, the unique properties of gallium arsenide (a rival to silicon) and the manufacture of silicon chips. The next three chapters describe the way electronic building blocks are used in control, instrumentation and telecommunications systems. Finally, Chapter 16 describes a simple system for assembling seven working circuits. I have included a handy glossary at the end of the book as well as some web addresses for taking your studies further.

Malcolm Plant, 2010

Only got a minute?

In one way or another we are all users of electronics. Every time you phone for a taxi, watch TV or send an email you are depending on the creativity of designers of electronic circuits and systems. Yet so much of what they create is hidden away from us. After all, how many of us can claim to have come face to face with a silicon chip or a communications satellite? A silicon chip is so small that you would lose it under your finger nail, and a communications satellite is somewhere in orbit way above us!

Many people have little interest in knowing how electronics works; others, and perhaps you are one of them, want to understand how underlying concepts help to explain practical applications of electronics. This book begins with an overview of how electronics has come to change our lives; it explores the properties of components such as resistors, capacitors, diodes and transistors, and explains how they are used in the design of electronic switches, timers, oscillators, amplifiers and decision-making circuits. Three chapters explore the way these functions contribute to the design of instrumentation, control and telecommunications systems. Of course, you'll meet some basic maths on the way but that's essential if you want to understand how circuits work. Although this is not intended as a project book, the final chapter will show you how easy it is to build a working project without the need for soldering.

A theme running through the book is the crossing point between our analogue environment and the digital environment we have created through computer-based technology. Compared with the attractions of the former, the latter seems a rather dreary place as it processes information in binary numbers, 1 and 0. However, the computer can be programmed to enhance our creativity and enlarge our understanding of the non-technical analogue world.

5 Only got five minutes?

Life would be quite different without electronics. As Chapter 1 explains, electronics has an impact on almost everything we do: from toasters to telecommunications, from computer games to control systems, there's hardly a social activity where electronics is not involved in some way. So how does this book reflect this wide range of applications? Taking you back about 300 years, let's see what Pope was implying in the following lines which were published in 1713.

> *Not Chaos-like, together crushed and bruised,*
> *But, as the world, harmoniously confused:*
> *Where order in variety we see.*
> *And where, though all things differ, all agree.*

(Alexander Pope: 'Windsor Forest')

Pope was making an observation about the flora in Windsor Forest that, as an analogy, suggests a way of approaching your study of electronics. An amateur botanist (as Pope surely was!) might start by recognizing the similarities, rather than the differences, between the constituents of Windsor Forest. Thus, at first glance, an oak tree bears little resemblance to a bramble. But both have branches, leaves and flowers, and both need sunlight and water to survive. Similarly, a television looks quite different from a radio-controlled model aircraft but both have aerials, amplifiers and switches, and both need electrical power and a transmitter to be of any use. This 'order in variety' recognized by Pope is a useful model for learning electronics since it focuses attention on how a set of basic electronic functions can be interconnected in different ways to produce the electronic system you want. In this book I emphasize the purpose and function of circuit building blocks such as timers and oscillators (Chapter 6), rectifiers (Chapter 7), amplifiers (Chapter 8), logic gates (Chapter 9), and counters (Chapter 10) that, in various combinations, make up the many electronic systems that are in use today. Chapter 6 describes an integrated circuit known as a '555 timer', which functions either as a timer (a monostable) or as an oscillator (an astable) depending on the configuration of resistors and capacitors connected externally to this integrated circuit. In other words the 555 timer is regarded as a black box. An input–process–output model is used to show how this and other black boxes are linked together to produce an electronic system such as a thermometer (Chapter 4).

Some people find the systems approach unsatisfactory since they want to know *how* circuits work *inside* a black box before they design an electronic system by interconnecting black boxes. That's not to say that the inner workings of a transistor or integrated circuit are not interesting but you can easily get bogged down in design detail and never get close to the excitement of what electronics can achieve. My advice to you is to get into the systems mode of thinking and leave the technical details about how a device works to the specialists who make the devices!

As you learn how components are interconnected to produce useful circuits, reflect on the relationship between the digital and analogue worlds. Indeed, Chapters 13 and 14 are largely concerned with the circuits needed for these two worlds to speak to each other. By analogue I mean the non-technical real world where our senses respond to continuous changes in, for example, ambient temperature, sunlight and atmospheric pressure. On the other hand, there is the human-constructed digital world of the computer where qualities are measured in patterns of 1s and 0s. However, if we look more deeply into the analogue world, events and processes seem less continuous. For example, look closely at the minute hand of a mechanical analogue watch and you will notice that the hand moves forward in small steps. This is because the watch has an escapement mechanism which determines the passage of time as ticks. Furthermore, quantum physics tells us that light radiation is composed of small packets of energy called photons. Is reality at the level of atoms a digital world and does it become smoothed out in our everyday analogue world?

1

....

Electronics today and yesterday

In this chapter you will learn:

- *about several appliances or gadgets that make use of electronics*
- *how electronics contributes to people's enjoyment of a modern way of life*
- *about the main developments in electronics over the years*
- *to consider the ethical implications of using electronics.*

1.1 The electronic age

Electronics is the art of using electrons in devices such as transistors and silicon chips to make electricity work for us. There is no doubt that this 'art' has had far-reaching effects on nearly all aspects of life, although its influence on everyday life often remains unseen. Our modern-day dependency on electronics was spurred by the invention of the transistor in the late 1940s. This was followed by the manufacture of silicon chips in the early 1960s that has ultimately led to incredibly complex circuits containing thousands of transistors integrated on a sliver of silicon so small that you could lose it under your fingernail. Miniaturizing electronic circuits in this way is called microelectronics and it has come to influence the way we store, process and distribute information; to change the way we design and manufacture industrial goods; to improve the diagnosis and treatment of illnesses; and it shapes the affairs of finance and business, as well as a variety of social, educational and political activities. Nowadays, we take for granted the way electronics makes our lives more comfortable, enjoyable, creative and exciting, so let us begin the study of electronics by taking a closer look at some of its applications.

1.2 Consumer electronics

We are surrounded by products in and around our homes that make use of electronics in one way or another. Washing machines, security systems and toasters are 'clever' because of the function of electronic circuits in them, not forgetting that hi-fi systems, radios, DVD (digital versatile disc or *disk*) players, GPS automotive navigation systems, MP 3 players and computers are so obviously 'electronic' in what they do. One major advance familiar to all is the storage of recorded information brought about by the CD (compact disc) and the DVD. A CD is a small and portable disc made of moulded polymer for electronically recording, storing and playing back audio, video, text and other information stored in digital form as microscopically small pits along a set of concentric tracks that spiral outwards from the centre of the disc. Each pit is about a thousandth of a millimetre long and about a tenth as deep. The track is so narrow that thirty tracks are about as wide as a human hair. Sixty minutes of sound recording requires about ten million such pits. The disc is rotated at high speed in a player in which a finely focused laser beam 'reads' the information on the disc. The stream of on/off pulses of laser light is converted to electrical signals that are processed to produce sound, say, that is almost the free from distortion (e.g. surface hiss) that can spoil the sound produced by the rather old-fashioned record player using a vinyl disc. Initially, CDs were read-only, but newer technology allows music lovers to record audio files, or other data, as well.

The DVD is also a form of CD that stores much more information in the same space, and can hold enough data for a 133-minute movie.

One of the most important developments in music technology has been the creation of the MP3 format for recording and playing music in digital form. MP3 is a compression system for music that reduces the amount of memory space required to store a song. The aim of this format is to compress a CD-quality song by a factor of 10 to 14 without noticeably affecting the quality of the sound. Whilst this is acceptable to most listeners, it is not considered sufficiently high fidelity by audiophiles. However, with MP3 you can compress down to about 3-megabyte (MB) a 32 MB song which enables you to download a song in minutes rather than hours, and store hundreds of songs on your computer's hard drive or on an MP3 player without taking up that much space.

Figure 1.1 A Sony Walkman mp3 player X series.

Courtesy: Sony

The MP3 compression system was designed by the Moving Picture Experts Group (MPEG) which was formed by several teams of engineers in Germany and the United States.

MP3 provides a representation of sound by discarding or reducing the precision of audio components less audible to human hearing, and recording the remaining information in an efficient manner. This is similar to the principles used by JPEG, an image compression format familiar to owners of digital cameras.

I am sure that cameras that capture images on a film will be around for a long time yet, but many photographers are finding digital cameras more exciting and convenient to use. The so-called digital revolution has transformed the capture and processing of still and moving pictures. No longer is it necessary to process film to recover the image. In its place, at the focal plane of the camera lens is an electronic sensor called a charge-coupled device (CCD). The CCD converts the image produced by the camera

Figure 1.2 The Sony Cyber-shot digital camera model WX1.

Courtesy: Sony

lens into a digital image that is stored in an on-board integrated circuit as a digital file. In this state, the image can be viewed at any time and then downloaded for image processing on a computer for manipulating and printing or for use on Web pages.

1.3 Communications electronics

It is difficult to ignore the ubiquitous mobile phone that has brought about such remarkable changes in people's daily lives, although when used carelessly in public places it is often associated with annoyance to others. Energized by people's need to communicate with others whether for business or pleasure, the development of the mobile phone has been rapid and widespread enabling us to keep in touch with others while on the move almost anywhere on the planet. Moreover, the demand for Internet access and for other services via the 'mobile' is stimulating a wide range of facilities for this modern personal communication device that gets progressively 'smarter'.

When we use a mobile phone to make an international call, our message is more than likely forwarded to a recipient by a communications satellite in orbit round the Earth. Countries that can afford to make the powerful rockets necessary to launch communications satellites are often in the business of launching satellites for other nations. Europe has such a launch rocket, called *Ariane*. For example, in 2009 an *Ariane* rocket was launched from the European Space Agency launch centre in Kourou, French Guiana carrying a telecommunications satellite called HOT BIRD 10 to provide television and mobile phone coverage for Europe, the Middle East and North Africa. Many of these communications satellites are placed in what is termed a geostationary orbit. This is an orbit approximately 36,000 kilometres above the Earth's surface and taking 24 hours to orbit the Earth, the same as the Earth's daily rotation period.

Figure 1.3 Ariane 5's July 1 2009 launch carrying the communications satellite TerreStar-1.

Courtesy: ESA-CNES-Arianespace

It therefore remains stationary above a position on the Earth's surface so it never loses contact with receiving aerials on the ground. A single geostationary satellite can 'see' approximately 40 percent of the Earth's surface. Three such satellites, spaced at equal intervals of 120 degrees apart provide coverage of the entire globe.

Insight

Like the ubiquitous mobile phone, GPS (global positioning system) receivers have revolutionized global communication. GPS is a global navigation satellite system (GNSS) developed by the US Department of Defense. Anyone can use it anywhere. It uses between 24 and 32 Earth-orbiting satellites that transmit precise radio signals so that GPS users can find their current position, time and speed. People use it for navigation purposes in their cars or for walking across country, for example.

1.4 Computer electronics

Compared with the first computers of the 1940s, today's computers show how remarkable the advances in electronics have been. Power-hungry, room-sized and unreliable, these early computers have been replaced by a variety of compact, efficient and versatile computers such as laptops. Computers are essential for the reliable and safe operation of navigation systems and landing systems for aircraft, and for automatic motorway toll stations which provide revenue for new roads and environmentally friendly traffic flow systems. A computer loaded with the appropriate software can meet anyone's needs: the tourist wishing to create a video of a holiday in the Caribbean; an astronomer interested in studying light from a distant galaxy; an engineer planning a new motorway; an accountant working out the cost of running a business; all these and more make the computer a powerful tool.

As with so many other products and systems, computers owe their efficiency and compactness to the development of the silicon chip (also, simply, microchip or, more formally, integrated circuit). This is a small piece of silicon on which incredibly complex yet tiny miniaturized circuits are integrated by photographic and chemical processes as explained in Chapter 12. Perhaps the best known of these chips are the Pentium microprocessors made by Intel and the Power PC microprocessors developed by Apple, Motorola, and IBM. The microprocessor, one or more of which is the 'brain' of a computer, is the most complex silicon chip made today.

1.5 Control electronics

Electronics is an essential part of modern control systems. For example, anti-skid braking systems (ABS) on some cars depend on electronic devices that ensure that wheels do not lock and cause a skid. Safety in passenger and military aircraft relies on complex control systems. Moreover, the regulation of temperature and pressure in chemical factories and nuclear power stations depends on electronic control systems for the safe and efficient operation of the plant. In the home, the power of electric drills and food mixers can be controlled at the turn of a knob. In the greenhouse, automatic control of temperature and humidity is possible. Under the control of a program, the microprocessor in a washing machine instructs devices such as motors, heaters and valves to carry out a prearranged sequence of washing processes.

Robots are computer-controlled machines for which electronics finds important application. They are used for tasks that are repetitive (and often boring for humans) such as welding and painting, and fitting of car windows in a manufacturing plant. In places such as nuclear reactors and interplanetary space, computer-controlled robots can work where it would be hazardous for humans to do so. However, sometimes, making robots is an enthusiast's hobby!

Insight

The idea that human intelligence can be described so precisely that it can be replicated by a machine raises philosophical issues about the nature of the mind. These issues have been the subject of myth and science fiction for ages. For example, Mary Shelley's *Frankenstein* considers a key challenge for artificial intelligence: if a machine possesses intelligence, can it also feel, and if it can feel, does it have the same rights as a human being? This issue is known as 'robot rights' and is currently being considered by California's Institute for the Future.

1.6 Medical electronics

Electronic equipment is widely used to diagnose the cause of illnesses, to help nurses care for patients and to be of assistance to surgeons carrying out operations. For example, the electrocardiograph (ECG) is an electronic instrument that can help in diagnosing heart defects. By means of electrodes attached to the body, the ECG picks up and analyses the electrical signals generated by the heart. The ECG is just one of a number of instruments available to nurses and doctors in hospitals. A complete patient-monitoring system also records body temperature, blood pressure and heart rate and warns staff should there be any critical change.

The ultrasonic scanner is a computerized medical instrument for providing images of the body's internal organs. It uses high-frequency sound waves generated by an ultrasonic device held in contact with the body. A computer processes the echo pattern of the ultrasonic waves bouncing from inside the body to give an image on a screen. These ultrasonic waves are much safer than X-rays for examining the developing baby in the womb.

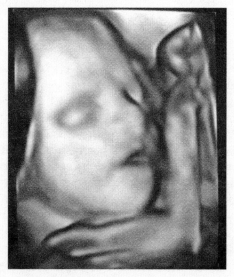

Figure 1.4 An ultrasonic scan of the unborn Daniel, son of family friends.

Now that microelectronics devices are so small and require very little electrical power to operate them, there are new challenges for bioelectronics. Bioelectronics has to do with implanting electronic devices in the body to help correct defects and so help people lead fuller lives. A long-term goal of medicine is the manufacture of a biochip – a silicon chip that is implanted in the body to make an on-the-spot check of chemical activity. For example, a biochip designed to respond to adrenalin in the blood could operate alongside a heart pacemaker. The pulse of the pacemaker could then adapt to varying degrees of excitement as the normal heart does. However, what if a biochip could be implanted in the brain – a sort of 'bolt-on intelligence'? This is not as far-fetched as you might think.

Researchers in Germany managed to affix multiple brain cells onto tiny transistor chips to create a part mechanical, part living electronic circuit that appeared to demonstrate that the cells communicated with each other. In this way, combining the mechanical abilities of electronic circuits with the extraordinary complexity and intelligence of the human brain, electronics may one day help the blind to see and the paralysed to move objects with their thoughts, and also help to build computers that are as inventive and adaptable as our own nervous systems, and a generation of robots that might truly deserve to be called intelligent. Could this be the first step to creating the humanoid android, *Data*, who is an indispensable crew member on the Starship *USS Enterprise*? The anticipation of such developments appears to turn science fiction into science fact and raises ethical issues about the way technology can blur the distinction between what is natural and what is artificial.

1.7 Looking back

More than a century ago, electronics was unknown. There were no radios, televisions, computers, robots or artificial satellites, and none of the products and services described above that we now take for granted. This revolution came naturally out of a study of electricity, familiar to the Greeks over 2000 years ago. Electricity was a subject of great interest to Victorian scientists, and to Sir William Crookes in particular.

1.8 The discovery of cathode rays

The beginnings of electronics can be traced back to the discovery of cathode rays in the closing years of the 19th century. These mysterious rays had been seen when an electrical discharge took place between two electrodes in a glass tube from which most of the air had been removed. Sir William Crookes called these 'cathode rays' since they seemed to start at the negative electrode (the cathode) and moved towards the positive electrode (the anode).

At that time, nobody had any idea what cathode rays really were. Nevertheless, during a historic lecture at the Royal Institution in London in April 1897, Sir J. J. Thomson declared that cathode rays were actually small, rapidly moving electrical charges. Later these charges were called electrons after the Greek word for amber.

Amber is fossilized resin from trees and has strange properties as the ancient Greeks had found. If rubbed with fur or a dry cloth, it has the power to attract small pieces of dust and fluff. Neither the Greeks, nor the scientists who devoted so much time to studying its properties in the period from the 17th century onwards, had a successful explanation of why amber behaved in this way. However, the discovery of the electron provided the answer. We now know that the electrical behaviour of amber (and of many other electrical insulators) is caused by static electricity. The friction between the cloth and amber causes electrons to be transferred from the cloth to the amber where they stay put to give amber an overall negative charge. This negative charge causes the amber to attract small bits of material to it.

1.9 The invention of the thermionic valve

The first practical application of cathode rays was the invention of the thermionic valve by Sir John A. Fleming in 1904. In this device, the heating of a wire (the filament) in an evacuated glass bulb produces electrons. The word 'thermionic' comes from 'therm' meaning 'heat', and 'ion' meaning 'charged particle', i.e. the electron. In a valve, negatively charged electrons driven out from the heated filament (the cathode) moved rapidly to a more positive anode. The flow of electrons stops if the anode becomes more negative than the cathode. This electronic component is called a diode since it has two (so 'di') electrodes for making connections to an external circuit. In addition, it acts like a valve because electrons flow through it only in one direction, from the cathode to the anode, not in the opposite direction.

It did not take long for an American, Lee de Forest, to make a much more interesting and useful thermionic valve. By adding a third electrode made of a mesh of fine wire through which the electrons could pass, he produced a triode. By adjusting the voltage on this third electrode (called the grid), he was able to make the triode behave like a switch and, more importantly, as an amplifier of weak signals. The triode made it possible to communicate over long distances by radio.

Insight

The development of the triode valve was demonstrated dramatically in 1912 when the luxury liner *Titanic* collided with an iceberg in the Atlantic Ocean. As this 'unsinkable' liner was going down, her radio operator broadcast an SOS radio signal using Morse code (dot-dot-dot, dash-dash-dash, dot-dot-dot) that was picked up by ships in the area, some of which went to rescue Titanic's passengers.

1.10 The beginnings of radio and television

Strangely, the First World War (1914–18) did little to stimulate applications for thermionic valves. But immediately after the war, electronics received a push that has gained strength ever since. In London the British Broadcasting Corporation was formed, and in 1922 its transmitter (call sign 2L0) went on the air. Firms such as Marconi, HMV and Echo made radio sets from components and valves supplied by Mazda, Ozram, Brimar and others.

The second major boost to the emerging electronics industry was the start of regular television transmissions from Alexandra Palace in London in 1936. But at that time the public had little interest in television; that was hardly surprising, as the pictures produced by John Logie Baird's mechanical scanning system were not very clear. By the time EMI had developed an electronic scanning system that gave much better pictures, the Second World War had begun. The Alexandra Palace transmitter was closed down abruptly in September 1939 at the end of a Mickey Mouse film for fear that Germany might use the transmission as a homing beacon for its aircraft to bomb London.

1.11 Radar and the Second World War

During the Second World War (1939–45) there were important advances in electronics. Perhaps the most significant invention was radar, developed in Britain to locate enemy aircraft and ships. The word radar is an acronym, for it is formed from the words **RA**dio **D**etection **A**nd **R**ange-finding. Radar was made possible by the invention of a high-power thermionic valve called the magnetron, a device nowadays commonly used as the source of microwaves in microwave cookers. In early radar, the magnetron produced high frequency pulses of radio energy that were reflected back from aircraft or ships to reveal their range and bearing.

Figure 1.5 A1920s Ethodyne superhet radio receiver.

Courtesy: The Science Museum

1.12 The invention of the transistor

In the period immediately following the Second World War, there was a major step forward in electronics brought about by the invention of the first working transistor. In 1948, Shockley, Barden and Brattain, working in the Bell Telephone Laboratories in the USA, demonstrated that a transistor could amplify electrical signals and also act like a switch. However, the way electricity moved in semiconductors, as these germanium-based devices were called, was not well understood. Furthermore, until the 1950s it was not possible to produce germanium with the high purity required to make useful transistors (see Chapter 2).

These transistors turned out to be successful rivals to the thermionic valve. They were cheaper to make since their manufacture could be automated. They were smaller, more rugged and had a longer life than valves, and they required less electrical power to work them. Once silicon began to replace germanium as the basic semiconductor for making transistors in the 1960s, it was clear that the valve could never compete with the transistor for reliability, compactness and low power consumption.

Consider ENIAC (another acronym, for Electronic Numerical Integrator And Calculator), the computer that was built in the 1940s at the University of Pennsylvania. ENIAC filled a room, used 18,000 valves, needed 200 kW of electrical power to work it, had a mass of 30 tonnes and cost a million dollars. The first transistorized desktop calculator of the 1960s was battery-powered, had a mass of a few kilograms and was capable of far more sophisticated calculations than ENIAC was able to perform. Moreover, this trend towards low-cost, yet more complex functions, to greater reliability yet lower power consumption continues to be an important characteristic of developments in electronics. ENIAC is now regarded as a first-generation computer, and the transistorized computers that followed it in the 1960s as second-generation computers. Third-generation computers required the development of the silicon chip.

Figure 1.6 The three generations of electronics: valve, transistor and integrated circuit.

1.13 Silicon chips make an impact

The first integrated circuits were made during the early 1960s. Techniques were developed for forming up to a few hundred transistors on a silicon chip and linking them together to produce a working circuit. The *Apollo* spacecraft that took men to the Moon in the late 1960s and early 1970s used these third-generation computers for navigation and control. The stimulus to miniaturize circuits in the form of integrated circuits came from three main areas: weapons technology, the 'space race' and commercial activity.

Modern weapons systems depend for their success on circuits that are small, light, quick to respond, reliable, and that use hardly any electrical power. Miniature circuits on silicon chips offer these advantages that had a profound impact on the space race which began when Russia launched Sputnik in 1957. At first, America's response was unsatisfactory, but she gained ground during the 1960s and Americans walked on the Moon by the end of the decade. Lacking the enormously powerful booster rockets developed by Russia, America needed compact and complex spacecraft and this stimulated the design of small and reliable control, communications and computer equipment. During the 1970s, spin-off from military interests and the space race further stimulated the growth of an electronics industry bent first on creating electronics goods and then on satisfying the demand for them in the home, the office and industry.

The 1970s saw the number of transistors integrated on a silicon chip doubling every couple of years and this trend continues as illustrated graphically in Chapter 12. Along with this increasing circuit complexity has been a similar doubling in the information processing power of the silicon chip. As noted above, the most important silicon chip is the microprocessor. It contains most of the components needed to operate as the central processor unit (CPU) of a computer. A highly complex device that can be programmed to do a variety of tasks, the microprocessor's versatility means that it acts as the 'brain' in a wide variety of devices. These fourth-generation computers have become faster and cheaper; they are now used in industrial robots and sewing machines, in space stations and toasters, in medical equipment and computer games. Their programmability and cheapness are their strength. The microprocessor brings the story to the present day but we can be sure that electronics will continue to influence our lives in ways that we can barely imagine.

TEST YOURSELF

1 *Name ten domestic appliances or gadgets that make use of electronics.*

2 *Describe three ways a computer enhances life at home or at work.*

3 *An electrocardiograph is:*
   (a) *a type of electronic game.*
   (b) *an instrument showing the electrical signals associated with heartbeats.*
   (c) *a device for stimulating muscles.*

4 *What is the name of the integrated circuit at the heart of a computer?*

5 *What can robots do more easily and safely than humans?*

6 *Write 100 words about the benefits and disadvantages of mobile phones.*

7 *A communications satellite is said to have a geostationary orbit when it:*
   (a) *stops the Earth rotating.*
   (b) *remains above one point of the Earth's surface.*
   (c) *sends signals to anywhere on Earth.*

8 *Describe three ways in which electronics is used for patient care in a hospital.*

9 *In 100 words, discuss the ethical implications of one development in electronics.*

10 *In 100 words, discuss the worries people have about accessing information on the Internet.*

2

....

The atomic roots of electronics

In this chapter you will learn:
- *about electrical insulators and electrical conductors*
- *that electrical resistance is a measure of the conducting properties of a material*
- *about the structure and properties of a neutrally charged atom*
- *the atomic structures of hydrogen, oxygen, carbon and silicon*
- *about the electrical properties of a semiconductor to compare insulators and conductors*
- *how p-type and n-type semiconductors are made by 'doping' pure silicon with so-called impurities.*

2.1 Electrons in atoms

In Chapter 1, I explained that electronics began with the discovery that cathode rays are actually beams of negatively charged particles called electrons. Nowadays, cathode rays are widely used to 'draw' information on television, computer and radar screens although this monopoly is being challenged by liquid crystal display (LCD) and light-emitting diode (LED) display technology. In cathode ray tubes, electrons are temporarily 'free' as they move through the vacuum inside, for example, a television tube or particular types of semiconductors. However, if electrons are the normal constituents of atoms, how can they become free? To answer this question requires a little knowledge of how atoms are structured.

Atoms are extremely small 'bits' of material – millions of them would lie side by side across the diameter of the dot at the end of this sentence. In what follows, it is convenient to regard the basic building blocks of atoms as comprising discrete 'bits' of material. At the core of an atom is a nucleus. Figure 2.1 shows a simple model of an atom in which atomic particles called protons and neutrons have their home in the nucleus. Electrons, however, make up the electron cloud outside the nucleus. The nucleus is very small compared with the overall size of an atom – say the size of an orange compared with the vast volume of an English cathedral such as Saint Paul's in London. Using this model of an atom, you can imagine the electrons to be a cloud of flies in the cathedral. Most of the mass of an atom lies in the nucleus. In fact, the neutron and proton have about equal masses whereas the mass of an electron is about 2000 times smaller than either particle.

Insight

Is it possible to see an atom? Well, not quite but the resolving power of the electron microscope is improving all the time so it may soon be possible to do so. The challenge is finding a wavelength of radiation that is smaller than the size of an atom with which to see the atom: that is, smaller than 10^{-10} metres. Whereas visible light is 10,000 times larger than this, X-rays with a much shorter wavelength offer a solution. If you want to know more about attempts at seeing the world of atoms, then this website will help: http://physicsworld.com/cws/article/print/23440

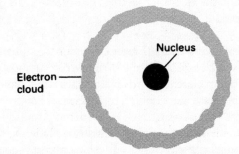

Figure 2.1 A simple model of an atom.

From the point of view of electronics, the two most important properties of an electron are its electrical charge and its small mass. Its electrical charge means that it can be moved by an electric field, as in a telephone wire, or between the electrodes in a cathode ray tube or through a semiconductor. Its small mass means that the path of a beam of electrons can be easily deflected by electric and magnetic fields as in a television tube or by small voltages as in a semiconductor. An electron carries a negative charge and the proton an equal magnitude but positive charge. Since these charges are of opposite sign, the electrons remain in an electron cloud around the nucleus, each electron occupying a particular energy level. The neutron does not carry an electrical charge, i.e. it is said to be electrically neutral, and it does not have any part to play in making electrons stay in the electronic cloud.

2.2 Atomic structure

Hydrogen and oxygen are two very common elements since their atoms go to make up a very useful liquid called water! Figure 2.2 shows that a hydrogen atom has the simplest structure of all atoms, since it has just one proton in its nucleus and one electron in the space surrounding the nucleus. This single proton identifies the atom as hydrogen since it is the number of protons in the nucleus of an atom that determines the physical and chemical properties of that atom. The electrical charge on a proton is equal and opposite to the charge on the electron, making the normal hydrogen atom electrically neutral.

An oxygen atom has a more complex structure as shown in Figure 2.3. Its nucleus is made up of eight protons and eight neutrons. Thus, an electrically neutral oxygen atom has eight electrons in the space surrounding the nucleus. Hydrogen and oxygen are just two of more than 100 different atoms in the universe, all made from the three main atomic building blocks, neutrons, protons and electrons. The table opposite summarizes the atomic structure of a few of these atoms.

Atom	Number of protons in nucleus	Number of neutrons in nucleus	Number of electrons in shells
Hydrogen	1	0	1
Oxygen	8	8	8
Copper	29	34	29
Silver	47	61	47
Silicon	14	14	14
Germanium	32	40	32
Carbon	6	6	6
Iron	26	30	26

Table 2.1 The atomic building blocks of some common elements.

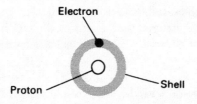

Figure 2.2 A hydrogen atom.

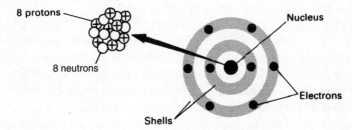

Figure 2.3 An oxygen atom.

2.3 Conductors, insulators and semiconductors

The reason why a particular material such as copper is a good electrical conductor is that it contains electrons that are bound quite weakly to the nuclei of the atoms of the material. These 'free' electrons can be moved easily by applying an electrical pressure, or voltage, between the ends of the material, e.g. from a battery. Copper is a good electrical conductor so it is used for connecting wires along which electrons flow easily between one device and another. Silver is a better conductor than copper but it is much more expensive, of course. Electrical insulators contain electrons that are more strongly bound to the parent nuclei and therefore free electrons are scarce. So, electrical insulators such as glass, polythene and mica are used to resist the flow of electrons between electronic devices, for example, as protective sheathing for wires and cables, to house electronic circuits, and so on. The property of a material that measures whether it is a good or bad conductor of electricity is known as its electrical resistance (Chapter 3).

Electronics has to do with the use of semiconductors as well as conductors and insulators. Semiconductors are the basis of electronics devices such as transistors and diodes, heat sensors called thermistors (Chapter 5), light-emitting diodes (Chapter 7) and integrated circuits (Chapters 12 and 13). As its name suggests, a semiconductor has an electrical resistance that falls somewhere between that of a conductor and an insulator.

Two of the commonest elements from which semiconductors are made are the chemical elements, silicon (symbol, Si) and germanium (symbol, Ge). Silicon is abundant in the Earth's crust and is present in sand and glass. Germanium was originally used for making transistors but it has been largely replaced by silicon except for specialist applications. Both elements are important in electronics because their resistance can be controlled to produce useful semiconductors. This is achieved by adding minute amounts of carefully selected substances to them, a process called **doping**.

We need to know something about the atomic structure of germanium and silicon to understand how this change comes about.

2.4 Silicon atoms

Figure 2.4 shows a model of a silicon atom that has 14 electrons surrounding a nucleus containing 14 protons and 14 neutrons. The part of this structure that makes silicon useful to electronics is the way that the electrons are arranged in what are known as shells surrounding the nucleus. There are two electrons in the inner shell, eight in the next shell and four in the outer shell. It is the four electrons in the outer shell, known as the valency shell, that make pure silicon a crystalline material.

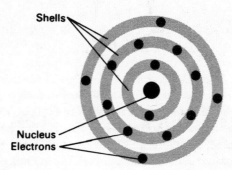

Figure 2.4 A silicon atom.

In a crystal of pure silicon, each of the four outer 'valence' electrons forms what is known as a covalent bond with an electron from a neighbouring silicon atom. Figure 2.5 shows how the pairing of electrons takes care of every one of these outer electrons. An orderly arrangement of silicon atoms results since there are no electrons that are unpaired so that pure silicon has a crystalline structure. Moreover, there are no free electrons available to make pure silicon conduct electricity and so it is an insulator. At least, it is an insulator at low temperatures, and a perfect insulator at the absolute zero of temperature (–273°C). However, we are not so much interested in how an increase of temperature reduces the resistance of silicon by breaking some of the covalent bonds, but in what happens to its resistance when a small amount of an 'impurity' is added to it.

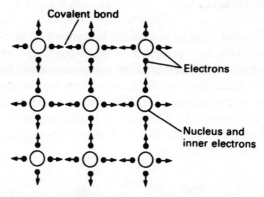

Figure 2.5 Pairing of electrons in silicon atoms.

2.5 *n*-type and *p*-type semiconductors

Once a very pure crystal of silicon has been manufactured, the silicon is doped with impurity atoms to make useful semiconductors! These impurity atoms are chosen so that they make a 'bad fit' in the crystal structure of silicon, due to the impurity atoms having too many or too few electrons in their outer shells. Depending on the impurity atoms, two types of semiconductor are produced in this way, n-type and p-type.

If silicon is doped with, for example, phosphorus atoms, an n-type semiconductor is produced. This arises because a phosphorus atom has five electrons in its outer shell. Figure 2.6 shows what happens when an atom of phosphorus is embedded in the crystal structure of pure silicon. Four of the five outer phosphorus electrons form covalent bonds with neighbouring silicon atoms, leaving a fifth unpaired electron. This unattached electron is now weakly bound to its parent phosphorus atom and it is therefore free to wander off and contribute to current flow through the semiconductor. Phosphorus is said to be a donor impurity since each atom of phosphorus can donate (give away) an electron. The addition of phosphorus has therefore changed the electrical properties of silicon. It has become an electrical conductor, not a good one but a *semi*conductor, due to the presence of free electrons donated by phosphorus atoms as shown in Figure 2.7. Thus, the 'n' in n-type semiconductor denotes the contribution that electrons make to current flow through this semiconductor and are known as majority charge carriers. There are also a very few electrons and holes (see below) produced by the effect of heat, which breaks covalent bonds between silicon atoms. The holes in an n-type semiconductor are called minority charge carriers. The contribution of minority charge carriers to the current that flows in n-type semiconductors is negligible, but note that the holes flow in the opposite direction to the electrons.

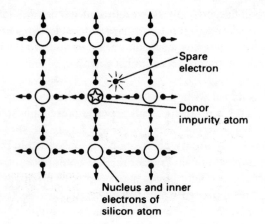

Figure 2.6 How a donor atom produces free electrons.

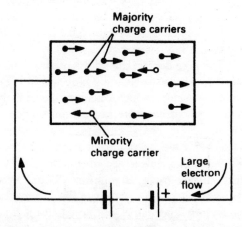

Figure 2.7 The flow of electrons in an n-type semiconductor.

By doping silicon with atoms such as boron that have three electrons in their outer shells, a p-type semiconductor is produced. Figure 2.8 shows what happens when a boron atom becomes embedded in the crystal structure of silicon. Three of its outer electrons become paired with neighbouring silicon atoms, leaving one unpaired silicon electron. Thus, there is no spare electron, as in an n-type semiconductor, but the vacant space will accept another electron. The vacancy created in silicon by doping it with boron is known (not surprisingly!) as a **hole**. Since this hole attracts an electron, it behaves as if it had a positive charge. Boron is said to be an acceptor impurity since it creates a vacancy in the crystal structure enabling it to accept an electron.

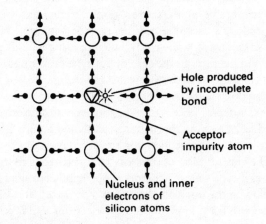

Figure 2.8 How an acceptor atom produces a hole.

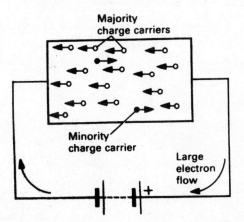

Figure 2.9 The flow of holes in a p-type semiconductor.

The presence of holes acting as positive charges in boron-doped silicon produces a p-type semiconductor ('p' for positive). In fact, current flow through a p-type semiconductor is most conveniently considered to be due to the movement of positively charged holes. The flow of holes can be likened to the movement of an empty seat in a row of theatre seats. If a seat at the end of a row becomes vacant (a hole), the person next to it might move into it and so produce a vacant seat. If the next person moved into the newly vacated seat (and so on), an empty seat would appear to move along the row. In the case of a p-type semiconductor, the people correspond to the electrons and the empty seat to the hole. This analogy helps us to understand that current flow in a p-type semiconductor arises from the movement of positive charge, the unfilled hole, moving through it as shown in Figure 2.9. The electrons in a p-type semiconductor are called minority charge carriers. The contribution of minority charge carriers to the current that flows in p-type semiconductors is negligible, but note that the electrons flow in the opposite direction to the holes. In Chapters 5 and 7 you will learn more about how electrons and holes in n-type and p-type semiconductors behave and how this behaviour accounts for the way diodes and transistors work.

TEST YOURSELF

1 State two examples of materials that are electrical insulators and two examples of materials that are electrical conductors.

2 Name the three main constituents of a neutral atom.

3 Sketch the atomic structures of hydrogen, oxygen, carbon and silicon.

4 Is the following statement true or false?

Free electrons must be present in a material if it is to conduct electricity easily.

5 An element widely used in the manufacture of semiconductors is silicon/carbon/iron?

6 Why is pure silicon an electrical insulator?

7 Explain how silicon becomes a p-type semiconductor when it is doped with boron.

8 Explain how silicon becomes an n-type semiconductor when it is doped with phosphorus.

9 Use the Internet to find out more about the origin of the term 'Silicon Valley'.

10 What is the most abundant element in the universe?

3

....

Simple circuits and switches

In this chapter you will learn:
- *to distinguish between series and parallel circuits*
- *how current, voltage and resistance are related*
- *how to do calculations involving powers of 10*
- *about Ohm's law*
- *the uses of different types of switch*
- *how to draw simple circuits to show how different switches are used*
- *how series and parallel circuits illustrate the function of simple decision-making logic gates.*

3.1 Making electrons move

Just as a gravitational force is required to make water flow downstream, an electrical force is required to make an electron move through a conductor. If the conductor is a copper wire as shown in Figure 3.1, electrons move through it when there is a difference of electrical force between its ends. This force is called a potential difference (p.d.) and is measured in volts (symbol V). Since a lamp lights when it is connected across the terminals of a battery, there must be a potential difference between these terminals to make current flow through the lamp. The electrical force, or potential difference, provided by a battery is known as an electromotive force (e.m.f.) (symbol E) of the battery and this is also measured in volts.

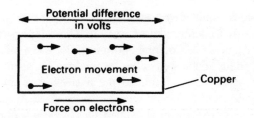

Figure 3.1 The flow of electrons through copper.

The flow of electrons through the lamp shown in Figure 3.2 is called an electric current (symbol I) and is measured in amperes (symbol A). A great many electrons are on the move in a conductor when a current of one ampere flows through it. In fact, about six million million million move past a point in the circuit each second! Since each electron carries an electrical charge, this current is a flow of electrical charge through the circuit. Electrical charge (symbol Q) is measured in coulombs (symbol C). When a current of one ampere flows through a circuit, the rate at which charge flows is one coulomb per second. Thus:

> 1 ampere = 1 coulomb per second
> or 1 A = 1 C/s
> or, in terms of quantities, $I = Q/t$

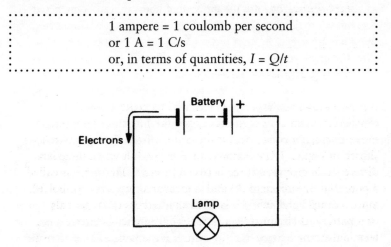

Figure 3.2 A simple circuit.

3.2 Series and parallel circuits

Devices such as lamps, switches, batteries and transistors are known as components. They are the individual items that are connected together to make a useful circuit. Figure 3.3 shows a simple circuit in which a battery, B_1, that has an e.m.f. of 6 V makes a current I flow through lamps L_1 and L_2. The switch SW_1 has two positions, open and closed. When the switch is closed, it offers a low resistance path and allows current to flow round the circuit. If SW_1 is open, it offers a high resistance path (the resistance of the air between the switch contacts) and stops current flowing through the circuit. The current comprises electrons that flow from the negative to the positive terminal of the battery. Before electrons were discovered, it was thought that electrical current flowed from the positive to the negative terminal of a battery. This direction is known as conventional current and is usually marked by an arrow on circuit diagrams. The circuit of Figure 3.3 is called a series circuit since the two lamps, the battery and the switch are connected one after the other. In a series circuit, the current flowing is the same at any point in the circuit so that the same current flows through each lamp.

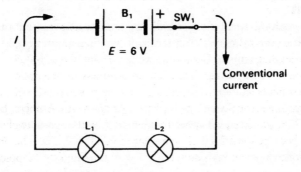

Figure 3.3 Two lamps connected in series.

Another common circuit arrangement of components that you will meet in your study of electronics is shown in Figure 3.4. In this circuit, two identical lamps, L_1 and L_2, are connected side-by-side to a battery, B_1, of e.m.f. six volts (6 V). Each lamp has the same p.d. across it, i.e. 6 V, since each is connected across the battery. When both switches are closed, the current flowing through each lamp is 0.06 A. So the total current provided by the battery is 0.12 A, the sum of the currents flowing through the two lamps. The circuit of Figure 3.4 is known as a parallel circuit. In this circuit, switch SW_1 independently controls lamp L_1, and SW_2 independently controls L_2. In the home, different appliances are connected in parallel with the mains supply so that each one is capable of being controlled by their respective on/off switches.

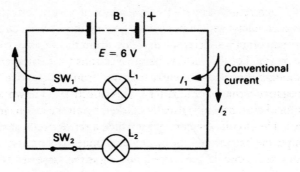

Figure 3.4 Two lamps connected in parallel.

3.3 Resistance and Ohm's law

Electrical resistance (symbol R) is a measure of the ease (or rather the difficulty!) with which electrical current is able to flow through a material. To remind you of the discussion in Chapter 2, copper has a low resistance since it is a good conductor of electricity; glass has a high resistance since it is a very poor conductor, i.e. an insulator. On the other hand, as explained in Chapter 2, p-type and n-type silicon have a resistance that falls somewhere between conductors and insulators and are known as semiconductors. The electrical resistance of a material is measured in units of ohms and is defined by the following equation:

$$\text{resistance} = \frac{\text{p.d. across the material}}{\text{current through the material}}$$

$$\text{or } R = \frac{V}{I}$$

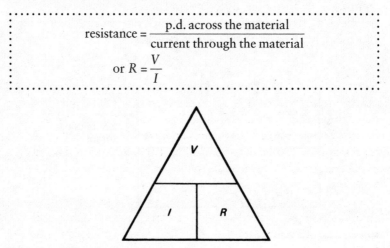

Figure 3.5 A triangle for working out V, I and R.

This equation can be used to find the resistance of one of the filament lamps shown in Figure 3.4. The p.d. across each lamp is 6 V, and is the same as the e.m.f. of the battery. The current through each lamp is 0.06 A. Thus, the resistance, R, of each lamp is found as follows:

$$R = \frac{6\text{ V}}{0.06\text{ A}} = 100\text{ ohms}$$

The unit of electrical resistance is the ohm; the symbol given to it is the Greek letter, Ω (omega). So the lamp has a resistance of 100 Ω. Of course, we could use the equation to find the current through a component, or the p.d. across a component. We would then need to use the above equation in a different form. Figure 3.5 helps to get the equations right.

To find R: cover R and $R = V/I$

To find V: cover V and $V = I \times R$

To find I: cover I and $I = V/R$

For example, suppose you want to find the current flowing through a 12 V car headlamp bulb that has resistance of 3 Ω. Since the current $I = V/R$,

$$I = \frac{12\text{ V}}{3\,\Omega} = 4\text{ A}$$

Note: The symbol, Ω, will be omitted from resistor values in all of the circuits in this book. Instead resistor values will be indicated by the multiplier 'R' for resistor values less than 999 ohms, by 'K' for resistor values between 1000 ohms and 99,999 ohms, and by 'M' for resistor values above and including 1,000,000 ohms. However, in the text, for clarity, the ohms symbol will be used.

If the resistance of a component is constant for a range of different values of potential difference and current, the component is said to be linear, or ohmic, and it obeys Ohm's law. Ohm's law is as follows:

Ohm's Law

Provided the temperature and other physical conditions of an electrical conductor remain unchanged, the potential difference across it is proportional to the current flowing through it.

Insight

Ohm's law, and the unit of resistance, honour the only major contribution to electricity made by the German physicist Georg Simon Ohm. Born in 1789, Ohm became a high school teacher where he developed an interest in the electrochemical cell, which was invented by the Italian Count Alessandro Volta. Ohm's experiments with home-made equipment led him to the conclusion that there is direct proportionality between the voltage applied across a conductor and the resultant electric current through it.

The graph in Figure 3.6a shows the behaviour of an ohmic device for which the ratio V/I is constant. In electronics, components called resistors obey Ohm's law quite closely. If the device is non-ohmic, then V/I is not a constant and it does not obey Ohm's law as the graph in Figure 3.6b shows. Non-ohmic conductors such as the thermistor and light-dependent resistor are described in Chapter 5.

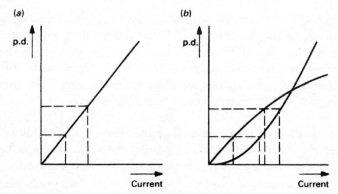

Figure 3.6 Comparison of ohmic and non-ohmic conductors. (a) Ohmic: if the p.d. doubles, the current doubles; (b) Non-ohmic: if the p.d. doubles the current is more than or less than doubled.

3.4 Large and small numbers

In electronic circuits, it is usual for values of current and potential difference to be small while values of resistance are large. For example, the value of a resistance might be one million ohms (1,000,000 Ω). Instead of writing down all the zeros for this large number, it is much easier to use the prefix 'M' meaning 'mega' for one million and write the resistance as 1 MΩ (or just 1 M on circuit diagrams). Similarly, small values of current can be expressed using the prefix 'm' meaning 'milli' for 'one thousandth of', or 'μ' meaning 'micro' for 'one millionth of'. Thus, ten milliamperes can be written as 10 mA, and 100 micro amperes as 100 μA. The table below summarizes the values of some of the prefixes you can expect to meet in working with electronic circuits.

Prefix	Factor	Powers of ten	Symbol
Tera	1 000 000 000 000	10^{12}	T
Giga	1 000 000 000	10^{9}	G
Mega	1 000 000	10^{6}	M
Kilo	1000	10^{3}	k
Milli	0.001	10^{-3}	m
Micro	0.000 001	10^{-6}	μ
Nano	0.000 000 001	10^{-9}	n
Pico	0.000 000 000 001	10^{-12}	p

Table 3.1 Large and small numbers.

When working with quantities in electronics that have large and small values, you will find it helpful to express the values as powers of ten rather than by writing down many zeros. The powers of ten equivalent to the prefixes are also listed in the table. Thus, the factor 1000 meaning 'one thousand times' is expressed as 10^{3}, meaning 'ten to the power three' ($10 \times 10 \times 10 = 1000$). In addition, the factor 0.000 001 meaning 'one millionth of' is expressed as 10^{-6}, meaning 'ten to the power minus 6' ($1/(10 \times 10 \times 10 \times 10 \times 10 \times 10)$). The numbers above the tens are called indices. Calculations become easier when large and small numbers are expressed as powers of ten because the indices can be added or subtracted.

For example, let us use numbers involving powers of ten in a sample calculation based on the equation $V = I \times R$. Suppose you want to work out the value of the potential difference across a component that has a resistance of 4700 ohms (4.7 kΩ). Suppose a current of two milliamperes (0.002 A, or 2 mA) flows through it. Now 4.7 kΩ can be written as 4.7×10^3 Ω ('four point seven times ten to the power three') where $10^3 = 10 \times 10 \times 10$. And 2 mA can be written as 2×10^{-3} mA ('two to the power minus 3') where $10^{-3} = 1/(10 \times 10 \times 10)$. Thus the equation gives

$$V = I \times R = 2 \text{ mA} \times 4.7 \text{ kΩ}$$
$$= 2 \times 10^{-3} \times 4.7 \times 10^3 \text{ volts}$$
$$= 9.4 \text{ V}$$

The calculation is simplified since the indices can be added (3 added to −3 equals zero). In addition, ten to the power zero, 10^0, equals 1.

Now take a second example. Suppose you need to know the current flowing through a component that has a resistance of 3.3 MΩ (3 300 000 Ω) when there is a potential difference of 4 V across it. Again, using the equation $I = V/R$, the resistance can be written as 3.3×10^6 Ω, so that $I = 4V/(3.3 \times 10^6)$ Ω. Since $1/10^6 = 10^6$, $I = 1.2 \times 1/10^{-6}$ amperes = 1.2 μA. Remember that $(1/1\,000\,000\ 1/10^6) = 0.000\,001$ or 10^{-6}.

3.5 Types of switch

Switches are used to turn current fully on or fully off in a circuit, not to some in-between value of current. A switch is 'on', or closed, when it offers a low resistance path for current flowing through it, and it is 'off', or open, if it offers a high resistance path to current flowing through it. We operate simple on/off switches dozens of times a day. A car's ignition, heater and radio are operated using on/off switches. So are cookers, hi-fi systems, televisions, radios and burglar alarms in the home. Moreover, the keyboard or keypad switches of calculators, computers and electronic games act to switch something on and off.

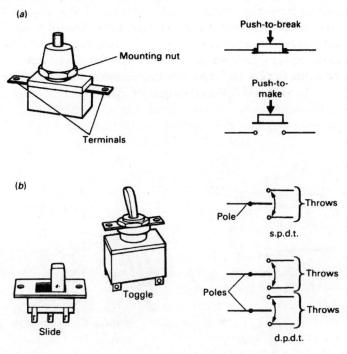

Figure 3.7 (a) Push-to-make, release-to-break switch. (b) Slide and toggle switches.

Many switches in everyday use require a mechanical force to operate them. The force brings together or separates electrically conducting metal contacts. Three types of mechanical switch are shown in Figure 3.7. The push-button switch (Figure 3.7a) is a simple push-to-make, release-to-break type. There are two circuit symbols for this type of switch depending on whether pushing makes or breaks the contacts. Slide and toggle switches (Figure 3.7b) are generally made either as single-pole, double-throw (s.p.d.t.), or as double-pole, double-throw (d.p.d.t.). The poles of these switches are the number of separate circuits that the switch will make or break simultaneously. Thus, a d.p.d.t switch can operate two separate circuits at the same time. An s.p.d.t switch is sometimes known as a change-over switch, since the pair of contacts that is made changes over as the switch is operated.

The microswitch shown in Figure 3.8a is simply a sensitive mechanical switch. It is usually fitted with a lever so that only a small force is required to operate it. This force causes contacts in the switch to open and close. The rotary switch shown in Figure 3.8b has one or more fixed contacts (its poles) that make contact with moveable contacts mounted on its spindle. Thus, a number of switching combinations can be made: e.g. 1-pole, 4-way; 2-pole, 6-way; 4-pole, 3-way and 6-pole, 2-way.

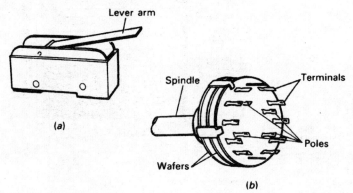

Figure 3.8 (a) Microswitch. (b) Rotary switch.

The two switches shown in Figure 3.9 are magnetically operated. The reed switch (Figure 3.9a) has two flat and flexible soft iron reeds that are easily magnetized and demagnetized. By sealing them inside a glass envelope, usually containing nitrogen (or some other chemically inert gas), the contacts are protected from corrosion. By bringing a permanent magnet close to a reed switch, magnetic induction causes the reeds to become temporarily magnetized so that they attract each other. The reeds make electrical contact, so closing a circuit. On removing the magnet, the reeds lose their magnetism and separate, so opening the circuit. The reed switch is a proximity switch, since it is operated by the nearness of the magnet. In addition, since the glass envelope protects the reeds, the reed switch is ideal for use in atmospheres containing explosive gases. Furthermore, it is a fast switch and can operate up to 2000 times per minute over a lifetime of more than a 1000 million switching operations.

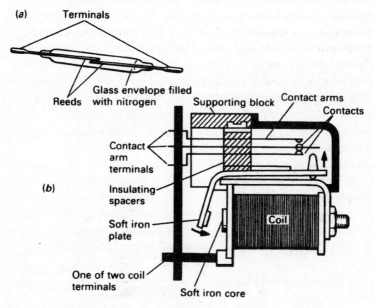

Figure 3.9 (a) A reed switch, and (b) an electromagnetic relay.

The second magnetically operated switch is the electromagnetic relay shown in Figure 3.9b. It has one or more pairs of contacts that are opened and closed by current flowing through an electromagnet consisting of insulated copper wire wound round an iron core. The current has the effect of magnetizing the iron and attracting a soft-iron plate to it, thus opening and closing the switch contacts at the end of the contact arms. The contacts are arranged as s.p.d.t or d.p.d.t, for example. The electromagnetic relay is a particularly useful switch for two reasons: the small current that energizes it enables a much larger current to be switched via its contacts; and the energizing current is completely isolated from the circuit that is switched on and off via its contacts.

In addition to the mechanical switches described above, electronics makes use of switches that have no moving parts. These are switches based on semiconductors. For example, the transistor is widely used in computer memories (Chapters 8 and 12), the thyristor that is used in the Rain Check project in Chapter 16, and the triac used for the control of power supplied to electric drills and food mixers (Chapter 7).

3.6 Simple digital circuits

Switches like the ones described above are on/off components. There is no in-between state enabling them to be in 'half on or half off', or 'nearly on or nearly off'. To explain the significance of this rather obvious statement, look at Figure 3.10a showing a simple circuit comprising a switch SW_1 connected in series with a battery and a lamp, L_1. When the switch is closed, the lamp is on; when it is opened, the lamp is off. Because there are just two states for this circuit, it is said to be a digital circuit. If we use the two

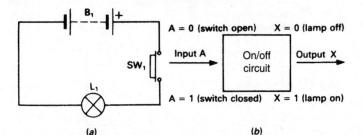

Figure 3.10 (a) A simple on/off circuit; (b) the circuit shown as a functional black box; (c) the truth table for this on/off circuit.

binary numbers, 1 and 0, to represent these two states, the number 1 represents the switch closed and the lamp on, and the number 0 represents the switch open and the lamp off.

The 'black box' drawn in Figure 3.10b is a symbolic way of showing the circuit. The two states of the switch represent the input information (1 or 0) to the box. In addition, the two states of the lamp, either on or off, represent the output information from the box. The table in Figure 3.11c summarizes the output and input information. The table is called a truth table since it 'tells the truth' about the function of the circuit.

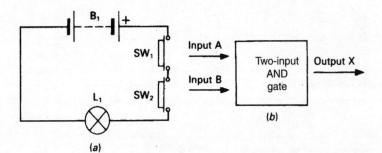

(a)

(b)

Input A (SW₁)	Input B (SW₂)	Output X
0	0	0
1	0	0
0	1	0
1	1	1

(c)

Figure 3.11 (a) A simple series circuit; (b) the circuit shown as a functional black box; (c) the truth table for this two-input AND gate.

Now look at the more complicated circuit shown in Figure 3.11a in which two switches, SW₁ and SW₂, are connected in series. Note that the lamp cannot light unless SW₁ *and* SW₂ are closed.

As the black box in Figure 3.11b shows, the switches provide input information and the lamp indicates the output information. The truth table in Figure 3.11c summarizes the function of this digital circuit known as an AND gate. It 'says' that the output state has a binary value of 1 (lamp on) only if switch SW_1 and switch SW_2 each have a value of binary 1 (both switches closed). If either or both of the switches are set to binary 0 (are open), the output is binary 0 (the lamp is off). Note that this digital circuit is called a gate because the switches open and close to control the information reaching the output just as a gate can be open or closed.

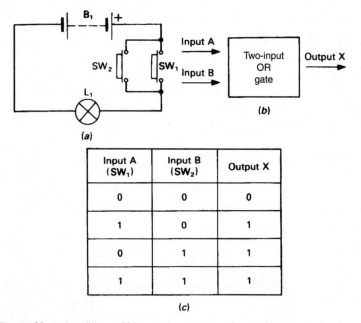

Input A (SW_1)	Input B (SW_2)	Output X
0	0	0
1	0	1
0	1	1
1	1	1

(c)

Figure 3.12 (a) A simple parallel circuit; (b) the circuit shown as a functional black box; (c) the truth table for this two-input OR gate.

A second simple digital circuit is shown in Figure 3.12a in which the two switches are connected in parallel. In this circuit the lamp lights if switch SW_1 or switch SW_2 is closed. The lamp also lights if both switches are closed. This OR gate has the truth table shown in Figure 3.12b that summarizes the values of the output information for all values of the input information.

The branch of electronics that is introduced by the above series and parallel circuits is known as digital logic. The AND and OR gates are called logic gates since their output states are the logical (i.e. predictable) result of a certain combination of input states. These logic gates, and others besides, are of great importance to electronics nowadays. Many thousands of logic gates are built from transistors in the form of integrated circuits that are used in a variety of equipment from control and communications equipment to calculators, watches and computers. You can learn more about logic gates in Chapter 9.

3.7 Using switches

CAUTION: The circuits below use a six-volt (6 V) battery as a low-voltage power source. Do not attempt to wire up similar circuits for mains operation unless you have the help and advice of a qualified electrician.

DOORBELL SWITCH (FIGURE 3.13)

This simple application of a single push-to-make, release-to-break switch (Figure 3.7a) enables a doorbell to be operated when the switch SW_1 is pressed.

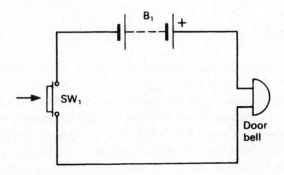

Figure 3.13 Doorbell switch circuit.

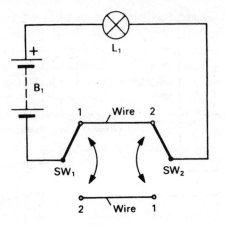

Figure 3.14 Two-way light switch circuit.

TWO-WAY LIGHT SWITCH (FIGURE 3.14)

Two single-pole, double-throw (s.p.d.t) switches (Figure 3.7b) are used in this circuit to enable a lamp to be switched on and off from two independent positions, e.g. from the top and bottom of a flight of stairs. In the position shown, switches SW_1 and SW_2 are in states that enable current to flow through the lamp. But note that SW_1 can be switched to state 2 isolating the lamp from the battery. Or SW_2 can be switched to state 1 to do the same thing. In either of these off positions, SW_1 may be switched to state 1, or SW_2 to state 2, to switch the lamp on again.

MOTOR REVERSING CIRCUIT (FIGURE 3.15)

This uses one double-pole, double-throw toggle (Figure 3.7b) to reverse the direction of rotation of the motor. The motor must be a direct current (d.c.) type; one that reverses direction when the current flows in the opposite direction through it. The switch is shown in both of two possible states. The arrows trace the path of the current from the battery via the switch contacts through the d.c. motor.

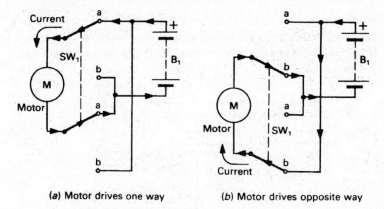

| **(a)** Motor drives one way | **(b)** Motor drives opposite way |

Figure 3.15 Motor reversing circuit (the broken line shows that the two parts of SW₁ operate together).

REVOLUTION COUNTER (FIGURE 3.16)

Here is a simple use of a microswitch (Figure 3.8a) to count the number of revolutions made by a wheel. Each time the wheel rotates a cam operates the arm on the microswitch. Since the microswitch has been fitted with a roller at the end of its lever arm, little frictional force is required to operate it.

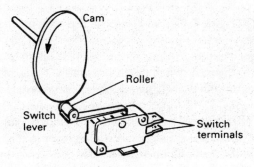

Figure 3.16 A revolution counter.

CHANNEL SELECTOR (FIGURE 3.17)

A single-pole, 12-way rotary switch (Figure 3.8b) can easily be used to select any one of 12 channels of information, e.g. radio stations 'piped' to hotel rooms, and selected by a rotary switch by the side of the bed, or the wavebands on a personal radio receiver. Rotation of the switch connects one of the incoming channels to the loudspeaker.

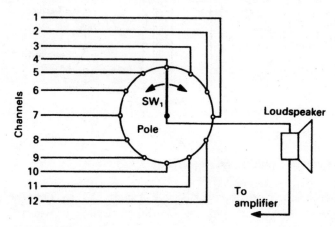

Figure 3.17 Channel selector circuit.

INTRUDER ALARM (FIGURE 3.18)

The reed switch (Figure 3.9a) can be used as a proximity switch to detect when a door or window is opened. Two small magnets are required, one inset in the door (it can be made invisible), and the other next to the reed switch that is inset in the frame of the door. The latter magnet ensures that the reeds on this reed switch are closed when the door is open so that the alarm sounds if the 'alarm-set' switch is closed. However, when the door is shut, the magnetic fields of the two magnets cancel each other in the region of the reeds and the reeds are open. Hence, the reeds close and the alarm sounds only when the door is opened. More than one door and window can be monitored in this way, merely by connecting all the reed switches in parallel with each other so that the closure of any reed switch sounds the alarm.

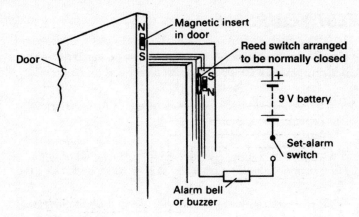

Figure 3.18 Intruder alarm circuit.

TEST YOURSELF

1 *Draw circuits showing two lamps in series connected to a battery, and two lamps in parallel connected to a battery.*

2 *In what unit is electric charge measured? How is electrical charge related to current?*

3 *A current of 4 A flows through a 12 V car headlamp bulb. How many coulombs pass through it in 10 seconds?*

4 *What is the resistance of the car headlamp bulb in question 3?*

5 *Name three types of mechanical switch.*

6 *Two identical switches are needed to switch a motor on and off from two different positions. What type of switch is required?*
   (a) *a single-pole, double throw, or*
   (b) *a double-pole, double throw?*

7 *Two students each have a single-pole, on-off switch. Draw a circuit that will enable a warning lamp to flash on when either switch is pressed.*

8 *What changes would you make to the answer to question 6 if the lamp is to come on only when both students press their switch?*

9 *A rocket being prepared for launch has two crew members, each of whom has a two-position switch, one position marked 'hold' and the other marked 'go'. Draw a circuit so that:*
   (a) *a 'hold' light comes on when either crew member has their switch in the 'hold' position.*
   (b) *a 'go' light comes on only if both crew members have their switch in the 'go' position.*

10 *Draw a circuit that lights a lamp when a window is opened. Your design for this intruder alarm should make use of two magnets and a normally open reed switch.*

4

....

Signals and systems

In this chapter you will learn:

- *the difference between direct current and alternating current*
- *what is meant by amplitude, frequency and period of a signal*
- *how to use ammeters and voltmeters in making measurements of current and voltage*
- *how an oscilloscope enables the properties, such as amplitude and frequency, of rapidly changing signals to be measured*
- *to design common electronic systems as interconnected sets of circuit building blocks.*

4.1 Direct current and alternating current

Electric current can be either direct current or alternating current. The type of current delivered by a battery in a torch is direct current (d.c.). In normal language we often say 'd.c. current' or 'a.c. current' although this really means saying 'current' twice! Though this d.c. current may vary in strength as when a battery wears out, the current does not change direction. Graphically, a direct current is shown in Figure 4.1a. As time passes, the current remains at a steady level. The current will eventually fall to zero but does not change direction. In a graph such as this, the plus (+) and minus (–) signs are used to indicate the two possible directions of current flowing through the circuit. For a d.c. current, the current flow is considered to flow from the positive terminal of the battery to the negative terminal and is called the conventional current flow. Figure 4.1b shows a varying direct current, one that would be produced if a switch was regularly opened and closed in a simple circuit. Closing the switch causes the current to rise abruptly to a maximum value where it remains steady until the switch is opened, whereupon it falls to zero.

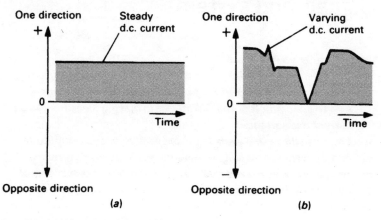

Figure 4.1 Graphs of (a) a steady d.c. current, and (b) a varying d.c. current.

An alternating current (a.c.) is one that flows first in one direction and then in the opposite direction. Figure 4.2a shows the variation in the alternating current from the mains supply. It smoothly increases to a maximum in one direction and then falls to zero before increasing to a maximum in the opposite direction. For the mains supply, this smooth and repetitive variation in current is called a sinusoidal waveform since the graph is the mathematical shape of a sine wave. Sometimes alternating electronic signals are in the form of a square wave as shown in Figure 4.2b. Whilst the current is shown on the diagrams, we might just as well have drawn them for the voltages that create the currents.

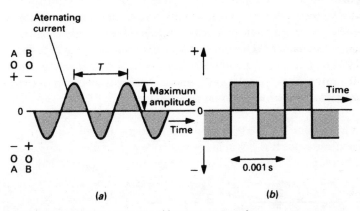

Figure 4.2 Graphs of (a) a sinusoidal a.c. waveform, and (b) a square wave a.c. waveform.

There are three significant characteristics of a regularly repeating
waveform of the type shown in Figure 4.2a. These are amplitude,
period and frequency. The amplitude is the value of the current
or voltage (or of any other quantity that varies in this way) at any
time. More often, we are interested in the maximum amplitude
reached by the waveform. For the domestic a.c. mains supply,
the maximum (or peak) value of the a.c. voltage is about 340 V,
not 240 V which is called its root mean square (r.m.s.) value. The
r.m.s. value is the equivalent d.c. voltage that provides the same
heating effect as the a.c. supply. The **period** and frequency of an
a.c. waveform are related to each other. The period is the time
between two consecutive maximum or minimum values. For the
a.c. mains supply, the period, T, is 1/50 of a second (0.02 s). The
frequency is the number of complete periods of the waveform in
one second. For the a.c. mains supply, the frequency, f, is equal to
50 hertz (50 Hz). The relationship between frequency and period
is the simple equation $f = 1/T$. So, what is the frequency of the a.c.
waveform shown in Figure 4.2b?

Electrical signals that have sinusoidal and rectangular waveforms are common in electronics. The signals delivered by a microphone in a recording studio comprise a mixture of sinusoidal waveforms. These are called audio frequency signals since they are detectable by the human ear. Their frequencies fall in the range 20 Hz to 20 kHz, but mature people are unlikely to hear frequencies clearly that are above about 10 kHz. Sounds having frequencies above 20 kHz are known as ultrasonic sounds that dogs, bats and other animals such as dolphins can hear. Sounds having frequencies below about 10 Hz are inaudible to humans and are known as infrasound. It is possible for someone to be killed by the vibrations induced in the body by high-intensity infrasound.

4.2 The multimeter

Circuit designers could not test and develop components and circuits without the use of multimeters and oscilloscopes. A multimeter of the type shown in Figure 4.3 combines in one portable, battery-operated instrument the facilities for measuring d.c. and a.c. current and voltage as well as resistance. These measurements are made on a number of selectable ranges covering both small and large values, e.g. ranges having maximum values of 10 mA and 1000 V. When this multimeter is used as an ammeter to measure current, it is placed in series with a component as shown in Figure 4.4a. The resistance of the ammeter must be small if it is to have a negligible effect on the value of the current being measured. However, when the multimeter is used as a voltmeter to measure volts, it is connected in parallel with a component as shown in Figure. 4.4b. Since a voltmeter should not disturb the p.d. being measured by drawing current from the component, it must have a very high resistance.

Figure 4.3 Two types of multimeter: left, digital; right, analogue.

Courtesy: Farnell Electronics

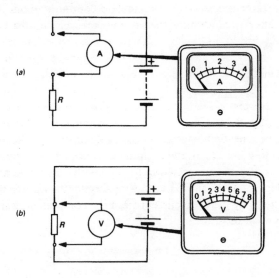

Figure 4.4 Using a multimeter (a) as an ammeter, and (b) as a voltmeter.

In an analogue multimeter of the moving coil type, this change of the resistance when switching the multimeter from 'volts' to 'amps' is achieved very simply. For the measurement of current, low value resistors are automatically connected in parallel with the meter terminals; this ensures that most of the current flows through this low resistance path leaving a calculated small amount to operate the meter movement. For the measurement of voltage, high value resistors are automatically connected in series with one of the meter terminals; this ensures that most of the applied voltage is 'dropped' across these resistors, leaving a calculated small amount to operate the meter movement.

When a multimeter is being used to measure the resistance of a component, it brings into action an internal battery, which makes a small current flow through the component. As shown in Figure 4.5, an ohmmeter produces a small current which flows through the component under test, and the display records this current. The smaller the current the larger the resistance of the component. This is why the resistance scale, in an analogue multimeter, has the zero of the resistance scale on the right of the display. What do you think is the resistance reading on the far left of the display?

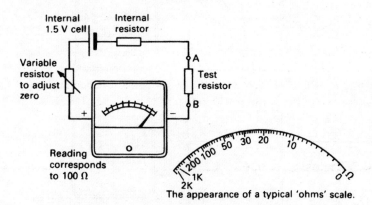

The appearance of a typical 'ohms' scale.

Figure 4.5 An analogue ohmmeter: (a) the internal circuit, and (b) a typical 'ohms' scale.

An analogue multimeter has moving parts comprising a coil of wire that twists in a strong magnetic field. The angle of twist is proportional to the strength of the current flowing through the coil. A pointer is attached to the coil and moves over a graduated scale. On the other hand, a digital multimeter does not have any moving parts, and what it measures is displayed on a liquid crystal display (LCD), or a light-emitting diode (LED) display. Digital multimeters are less likely to be damaged by mechanical shocks, and they incorporate 'user-friendly' facilities such as automatic ranging and a clear indication of the type of signal being measured. By plugging in sensors, a digital multimeter can be used to measure quantities such as frequency and temperature.

A digital multimeter performs rather better than an analogue multimeter when called upon to measure voltage. This is because a digital voltmeter draws less current from the circuit under test and therefore has less effect on the voltage being measured. This effect is given by the multimeter's sensitivity. Thus a general-purpose analogue multimeter might have a sensitivity of 20 kΩ per volt (20 kΩ V^{-1}), or 100 kΩ V^{-1}, while a general-purpose digital multimeter could have a sensitivity of at least 1 MΩ V^{-1}. The sensitivity is inversely related to the current drawn from the circuit under test. Thus, when a meter with a sensitivity of 20 kΩ V^{-1} measures a p.d. of 1 V, it draws a current of 1 V/20 kΩ or 50 μA from the circuit. This compares with 1 V/1 MΩ or 1 μA for the digital multimeter. However, remember that for a particular sensitivity, the higher the voltage range selected, the greater the resistance of the meter and the less it 'loads' the circuit under test. Thus, the 20 kΩ V^{-1} voltmeter has a resistance of 20 kΩ on the 1 V range and a resistance of 200 kΩ on the 10 V range.

4.3 The oscilloscope

The oscilloscope is a versatile instrument and it can be used to measure the characteristics of both d.c. voltages and a.c. voltages.

An oscilloscope like the one shown in Figure 4.6 can capture or 'freeze' on its screen part of a rapidly changing waveform so that measurements can be made of the frequency, shape and period of the waveform. It is therefore widely used in the design and development of amplifiers, music synthesizers, televisions, radios and computers. Oscilloscopes are generally operated from the mains power supply for use in laboratories and workshops. However, increasingly, electronic enthusiasts are using hand held oscilloscopes of the type shown in Figure 4.7. They are portable, lightweight and ideal for test and field use by engineers and scientists, especially those working outdoors. Rather than an electron gun they use a liquid crystal display. Some hand-held oscilloscopes combine the functions of a digital multimeter with the oscilloscope. Another type uses a laptop computer to process and display the characteristics of a signal. In this case, a specialised unit is connected to the USB port which provides signal isolation and incorporates an analogue-to-digital converter. (See Chapter 13.)

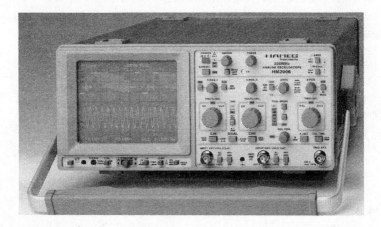

Figure 4.6 A cathode ray oscilloscope.

Courtesy: Farnell Electronics

Figure 4.7 A hand-held oscilloscope.

Courtesy: Velleman Instruments

The full name for the oscilloscope shown in Figure 4.6 is cathode-ray oscilloscope (CRO) since its main component is a cathode-ray tube (CRT). The CRT is also used in conventional televisions, visual display units and radar screens though they are rapidly being phased out in favour of LCD TVs. It incorporates an electron gun that 'fires' a narrow beam of rapidly moving electrons (cathode rays) at a phosphor-coated screen that glows at the point where the beam strikes it. The electron beam moves in sympathy with the signal waveform so it is possible to 'write' a graph (a trace) of the waveform on the screen. A dual-trace oscilloscope enables the shape of two waveforms to be written on the screen simultaneously. This is achieved by electronic circuits that move the electron beam rapidly between two separate parts of the screen thereby giving the impression of two independent signal traces on the screen. This enables the comparison of the shapes of two waveforms. Dual-beam oscilloscopes have two electron guns and they can produce the shape of two higher frequency waveforms simultaneously.

The inside surface of the screen of a cathode-ray tube is coated with a phosphor that lights up, or fluoresces, when electrons strike it.

Different phosphors produce different colours of light, though green is favoured for most 'scopes. Of course, once the phosphor has been activated by the impact of the electrons, its light must fade away within a few milliseconds if the trace is not to linger and confuse the picture. However, some cathode-ray tubes are designed to capture the shape of a single waveform so that it can be examined more easily. For example, where an electrocardiograph uses a cathode-ray tube, it retains the waveform of heart beats for a few seconds; and so does a radar screen that marks the position of aircraft by the 'echo' of radio waves that bounce off them.

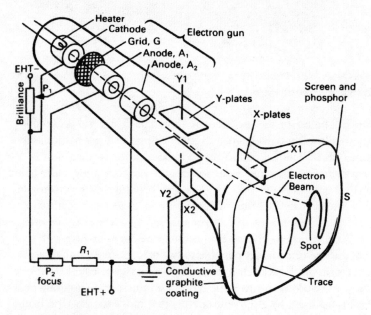

Figure 4.8 The operation of a cathode-ray tube.

Figure 4.8 shows the inner workings of the type of cathode-ray tube used in cathode-ray oscilloscopes. It has three main parts: an electron gun, a deflection system and a fluorescent screen, all housed in a glass envelope from which all the air has been taken out. This type of tube uses high voltages on the deflecting plates to change the path of the electron beam and to focus it into a small spot on the screen. This technique is called electrostatic deflection, as opposed to magnetic deflection that uses the magnetic field produced by a current passing through coils wound round the tube, used, for example, in most television picture tubes.

The electron gun comprises a heated tungsten filament within a nickel cathode cylinder coated with oxides of barium and strontium that give off electrons, a process known as thermionic emission, which was widely used in the (now almost obsolete) valve (see Chapter 1). The electrons are negatively charged so they accelerate towards the anodes A_1 and A_2 that are at a more positive voltage than the cathode. The strength of the electron beam, and hence the brightness of the image on the screen S, is controlled by the potentiometer P_1 making the grid more or less negative with respect to the cathode. The rest of the electron gun consists of accelerating and focusing anodes that are shaped metal cylinders, all held at a positive voltage that can be varied to alter the size of the spot produced on the screen. For a small CRT, the positive voltage on the focusing anodes is between 500 V and 1000 V. It is therefore dangerous to meddle with the circuits inside an oscilloscope – or a television, for that matter – unless you know what you are doing! A graphite coating inside the CRT avoids the build-up of electrical charge on the screen by collecting any secondary electrons given off by the screen.

After leaving the electron gun, the electron beam enters the deflection system that consists of two sets of metal plates, X and Y, at right angles to each other. A time-base circuit generates the voltage applied to the X-plates. Its job is to deflect the electron beam horizontally to make the spot sweep across the screen from left to right at a steady speed. The speed can be adjusted by the time-base controls on the oscilloscope. After each sweep, the time-base amplifier switches off the beam and sends it back to the starting point at the left of the screen – this process is known as flyback.

The waveform to be examined is amplified and applied to the Y-plates. The amplification of the waveform can be adjusted by the Y-sensitivity controls. The input waveform causes the horizontal trace to move vertically in response to the strength of the waveform. A stable trace appears on the screen when each horizontal sweep of the trace starts at the same point on the left of the screen. This is achieved by feeding part of the input waveform to a trigger circuit. This triggers the time-base circuit when the input signal reaches a magnitude that is set by the trigger level control. Most CROs allow manual and automatic adjustment of the triggering of the time-base.

In order to make measurements of the amplitude and frequency of a signal, there is usually a scale marked out as a grid of lines spaced at 10 mm (1 cm) intervals across the screen of a CRO. Thus, when a rectangular a.c. waveform is fed to the oscilloscope, the trace on the screen might appear as shown in Figure 4.9. Suppose this trace occurs when the Y-sensitivity control is set at 2 volts per cm (2 V cm^{-1}). Clearly, the amplitude of the signal is about 3 V. The frequency of the signal is obtained from the time-base setting. Suppose this setting is 1 millisecond per cm (1 ms cm^{-1}). Now the period of the signal is the distance marked T and this occupies a distance of 5 cm on the scale. This represents a time of 1 ms × 5 = 5 ms and since the frequency, f, of the waveform is equal to $1/T$, the signal has a frequency of 1/5 ms = 200 Hz. The scale also gives the mark-to-space (M/S) ratio of the signal. This is the ratio of the times, the signal is HIGH to when it is LOW (or a : b). From the scale, the M/S ratio is 3 : 2.

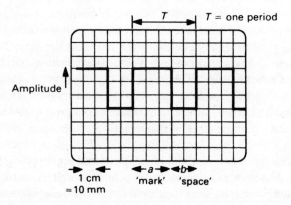

Figure 4.9 A typical trace on the screen of an oscilloscope.

4.4 Black boxes

Some people are more comfortable with learning *what* electronic components do rather than *how* they work. The advantage of doing this is that you avoid the possibility of getting so bogged down in the detail of how components work that you lose sight of the interesting ways they are used, in measurement, communications, control and computing, for example. From now on in this book, the emphasis will be on applications of electronics using the simple circuit principles discussed so far. So, let us take the example of an amplifier to show what we mean by this.

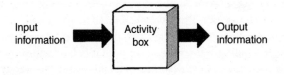

Figure 4.10 A black box amplifier.

Amplifiers can be found in radios, televisions, mobile phones and communications satellites – anywhere it is necessary to make the strength of a weak signal stronger. In order to understand the overall purpose of an amplifier, we simply need to know what effect the amplifier has on the properties of a signal that is fed into it. As shown in Figure 4.10, the amplifier is drawn as a 'black box'. It is called a black box not because of its colour (you can colour it any way you like!) but because we are interested in *what* it does, not how the circuits inside it do the job of amplifying. (Perhaps it is called 'black' because its contents are inscrutable?) The black box has a signal representing information going into it (called the input signal) that is modified by the black box in some way before coming out of it (called the output signal). In electronics, a black box is an activity box, since the properties of the signal coming out of it are different from the properties of the signal going into it. The light and heat sensors, digital counters, digital memories and optoelectronic displays described in the following chapters, are some of the many black boxes used in electronics. When black boxes such as these are connected together in a purposeful way, an electronic system is created. See Chapter 8 for more information about amplifiers.

4.5 Electronic systems

The thermometer shown in Figure 4.11 is an example of an electronic system of a thermometer for measuring temperature. It may be regarded as made up of three black boxes as shown by the systems diagram in Figure 4.12. The sensor (black box 1) is placed in contact with the item whose temperature you need to know. The signal produced by the sensor is input to the amplifier (black box 2). The amplifier produces an output signal suitable for operating the display (black box 3). A systems diagram like this simplifies a complex function, e.g. measuring temperature, and it helps to show what part each black box plays in the operation of the system even though you may not be sure how each black box achieves what it does.

Figure 4.11 A digital thermometer for use with K-type thermocouples (see Chapter 14).

Courtesy: Aemc (R) Instruments, Foxborough, MA, USA

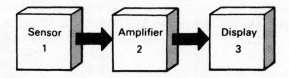

Figure 4.12 The electronic system of a thermometer.

A systems diagram for a radio receiver is shown in Figure 4.13. It comprises a tuned circuit (black box 1) that selects the narrow band of radio frequencies used to carry a message from a transmitting station. A radio frequency amplifier (black box 2) amplifies these radio frequencies. A detector and filter (black box 3) then converts the amplified frequencies into a form suitable for operating an earpiece. Further amplification by the audio amplifier (black box 4) enables a loudspeaker to be operated (black box 5).

Figure 4.13 The electronic system of a radio receiver.

Figure 4.14 shows the systems diagram for a computer. A microprocessor (black box 1) carries out a list of instructions held in a memory (black box 2). The windows to the outside world of the computer are the input ports and the output ports (black box 3) through which information comes and goes. Within the computer, information travels along electrical conductors called the 'highway' (or bus) that connects together the various parts of the computer system.

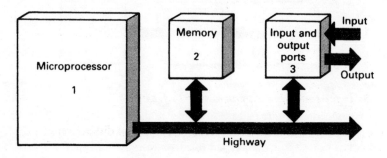

Figure 4.14 The electronic system of a computer.

4.6 Analogue and digital systems

The words analogue and digital are generally used for describing electronic devices such as the analogue and digital multimeter described above. For example, electronic watches are said to be analogue or digital according to the type of display they use. The word 'analogue' means 'model of', so an analogue watch models the smooth passage of time by using hands that move smoothly round its face. The advantage of this type of display is that it is easy to get an idea of the present time in relation to past and future time as represented by the numbers on the face of the watch. The word 'digital' means 'by numbers', so the digital watch displays the current time as a set of numbers that change abruptly at intervals. Some electronic devices combine both analogue and digital functions. Thus, an electronic analogue watch uses digital circuits for timing but displays the time in analogue form. It is interesting to note that spring-driven watches work digitally, in the sense that the escape mechanism monitors time as a series of 'ticks'. I wonder if time itself, the passage of time, is analogue or digital.

TEST YOURSELF

1 *Is the current provided by a 9 V battery a.c. or d.c.?*

2 *Is the mains domestic supply a.c. or d.c.?*

3 *Is the following statement true or false?*

A voltmeter is used for measuring potential difference.

4 *An ammeter measures electrical current, resistance or voltage?*

5 *Voltmeters are placed in .... with a component across which voltage is being measured.*
Series or parallel?

6 *A moving coil ammeter should have a .... resistance.*
Low, medium, high?

7 *Is the following statement true or false?*

The best voltmeters require a high current to operate them.

8 *Cathode rays are ....*
Rapidly moving molecules, a type of laser beam or a stream of electrons?

9 *An oscilloscope displays the waveform shown in Figure 4.15. if the Y-sensitivity is set at 2 v cm^{-1}, what is the amplitude of this waveform?*

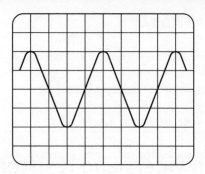

Figure 4.15 Question 9.

10 *An oscilloscope displays the waveform shown in Figure 4.16. If the time per cm is set at 2 ms per division, what is the frequency of the wave?*

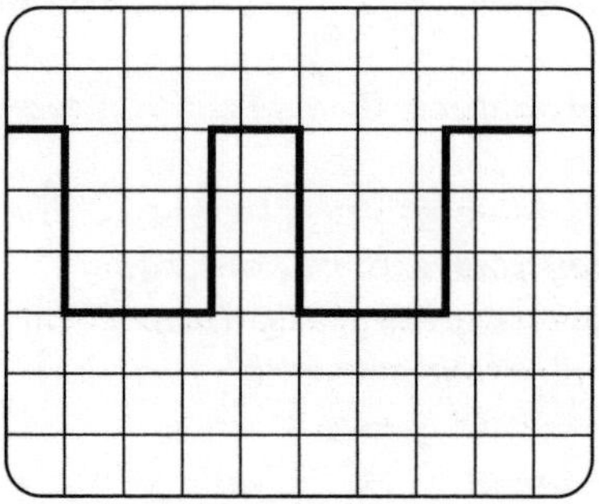

Figure 4.16 Question 10.

5

....

Voltage dividers and resistors

In this chapter you will learn:

- *to draw and explain the purpose of a voltage divider*
- *how to design a voltage divider to provide a particular output voltage*
- *to calculate the resulting values of resistors connected in series and parallel*
- *to identify fixed-value resistors from their colour coding*
- *how to select resistors for their power rating*
- *about the function and use of a light-dependent resistor, a thermistor and a strain gauge*
- *the properties and use of a strain gauge.*

5.1 The potential divider

Figure 5.1a shows the function of this useful circuit building block. It provides an output potential difference, V_{out}, that is less than the input potential difference, V_{in}. However, why should this be a useful function? Well, potential dividers are used as volume controls in radios, and for controlling the brightness of television screens, for example. As you will see, potential dividers are widely used in other circuit designs, e.g. in control and instrumentation systems where a voltage has to be reduced to a value suitable for operating transistors and integrated circuits, or for fixing a voltage at a preset level.

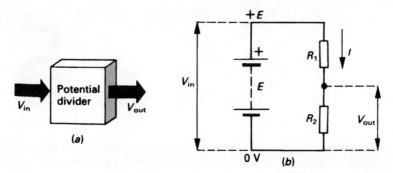

Figure 5.1 (a) The function of a potential divider. (b) The use of two resistors as a potential divider.

Figure 5.1b shows a circuit that acts as a potential divider by dividing the e.m.f. generated by a battery. The two rectangular symbols, marked R_1 and R_2, are electronic components called resistors; they simply have an electrical resistance measured in ohms. The reduced output voltage measured with respect to 0 V, occurs at the junction between the two resistors. It is possible to obtain any value of p.d. between 0 V and V_{in}, the e.m.f. of the battery, by changing the values of the resistors. The values of the resistors R_1 and R_2 determine the output p.d. V_{out}. The equation is

$$V_{out} = \frac{V_{in} \times R_2}{R_1 + R_2}$$

This equation shows that V_{out} is less than V_{in} by the fraction $R_2/(R_1 + R_2)$, i.e. the smaller R_2, the smaller V_{out}. Just suppose that $V_{in} = 9$ V and the values of the resistors are $R_1 = 90 \ \Omega$ and $R_2 = 10 \ \Omega$. Now

$$V_{out} = \frac{9 \text{ V} \times 10}{100} = 0.9 \text{ V}$$

The same value for V_{out} could have been obtained if $R_1 = 900 \ \Omega$ and $R_2 = 100 \ \Omega$, for then

$$V_{out} = \frac{9\ V \times 100}{1000} = 0.9\ V$$

Or if $R_1 = 240\ \Omega$ and $R_2 = 120\ \Omega$, then

$$V_{out} = \frac{9\ V \times 120}{360} = 3\ V.$$

Note that it is the ratio of the values, not the actual values of the resistors, that determines the output voltage of a potential divider. You can prove the above equation very simply by using the relationship between V, I and R that was given in Chapter 3. First note that the current, I, flows through both resistors, and is given by

$$I = \frac{V_{in}}{R_1 + R_2}$$

In addition, V_{out} is given by $I \times R_2$.

If we substitute I from the first equation into the second,

$$V_{out} = \frac{V_{in} \times R_2}{R_1 + R_2}$$

as required.

5.2 Resistors in series and parallel

The combined resistance of two resistors connected in series is found by adding together their values. Thus in Figure 5.2a the total resistance, R, of two resistors R_1 and R_2 connected in series is given by the equation

$$R = R_1 + R_2$$

This equation can be easily proved. First, note that when two resistors are connected in series the same current, I, flows through each resistor. Second, the sum of the p.d.s across the two resistors is equal to the p.d. across the combination. Thus $V = V_1 + V_2$. And since $V = I \times R$, we can write

$$V = I \times R_1 + I \times R_2 = I \times (R_1 + R_2) = I \times R$$

Here we have written $R = R_1 + R_2$ for the resistance of the combination. Thus, if we replace the two resistors connected in series by a single resistor equal to the sum of the combination, the current drawn from the battery remains unaltered.

As shown in Figure 5.2b, when two resistors are connected in parallel the total resistance of the combination is given by the equation

$$\frac{1}{R} = \frac{1}{R_1} + \frac{1}{R_2}$$

This can be rewritten

$$R = \frac{R_1 \times R_2}{R_1 + R_2}$$

This equation can be proved by first noting that the p.d. across each resistor is equal to the p.d. across the combination. Second, the sum of the currents through each resistor is equal to the current flowing from the power supply. Thus $I = I_1 + I_2$. And since $I = V/R$, and $I_1 = V/R_1$ and $I_2 = V/R_2$, we can write

$$I = \frac{V}{R} = \frac{V}{R_1} + \frac{V}{R_2}$$

This equation reduces to

$$\frac{1}{R} = \frac{1}{R_1} + \frac{1}{R_2}$$

Note that when two or more resistors are connected in series, their total resistance is more than the largest value present. On the other hand, when two or more resistors are connected in parallel, their total resistance is less than the smallest value present. We can prove this by taking two values for R_1 and R_2, e.g. let $R_1 = 300\ \Omega$ and $R_2 = 500\ \Omega$. If these two resistors are connected in series, their combined resistance is $300 + 500 = 800\ \Omega$, that is more than the largest value present, i.e. more than $500\ \Omega$. Additionally, if they are connected in parallel, their combined resistance is

$$\frac{300 \times 500}{300 + 500} = \frac{150000}{800} = 187.5\ \Omega$$

This is less than the smallest value present, i.e. less than $300\ \Omega$.

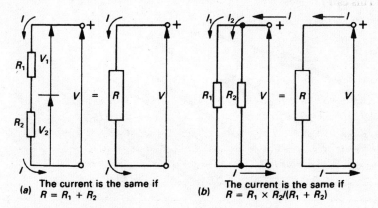

(a) The current is the same if $R = R_1 + R_2$

(b) The current is the same if $R = R_1 \times R_2 / (R_1 + R_2)$

Figure 5.2 (a) Two resistors connected in series, and (b) two resistors connected in parallel.

5.3 Fixed-value and variable resistors

Figure 5.3 shows examples of fixed-value resistors. By depositing a hard crystalline carbon film on the outside of a ceramic rod, a carbon-film resistor is made. It is then protected by means of a hardwearing and electrically insulating coating. The resistance of the carbon film between the connecting wires is the resistor's value. Of similar construction is the metal-film resistor, except that tin oxide replaces the carbon. The metal-film resistor has better temperature stability than the carbon-film resistor. Both types are recommended for use in audio amplifiers and radio receivers where they may be exposed to extreme changes of temperature and humidity. Metal-film resistors also generate little electrical 'noise'. (Electrical noise is the 'hiss' that can tend to 'drown' the required signal in a circuit, and is caused by the random movement of electrons in the resistor.) By winding a fine wire of nichrome (an alloy of nickel and chromium) round a ceramic rod, a wire-wound resistor is made. Wire-wound resistors can be made to have a very precise value, guaranteed to within 0.1 per cent.

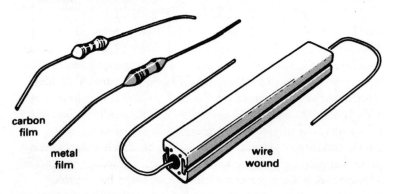

carbon
film

metal
film

wire
wound

Figure 5.3 Examples of fixed-value resistors.

The thick-film resistor is made by adjusting the thickness of a layer of semiconducting material to give the required resistance. These resistors are generally grouped eight at a time in a single-in-line or dual-in-line package (Figure 5.4). The individual resistors may be independent of each other or have a common connection, depending on the application. Thick-film resistors of this type are particularly useful in computer circuits where eight or more connections have to be made between the computer and display or control circuits.

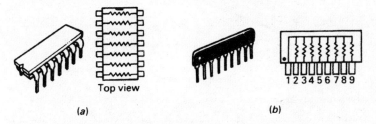

Figure 5.4 Resistors grouped together as (a) a dual-in-line (d.i.l.) package, and (b) a single-in-line (s.i.l.) package.

Figure 5.5 shows the two main types of variable resistor, together with their circuit symbols. They each have three terminals: two make contact with the ends of a carbon or wire-wound resistive track and the third is attached to a 'wiper' that moves over the track. The resistance between one end of the track and the wiper varies as the wiper moves along the track. The three terminals enable the variable resistor to act as a potential divider (section 5.1) since effectively it comprises two variable resistors connected in series. As the resistance of one resistor increases, the other decreases. When a variable resistor is used as a potential divider, it is known as a potentiometer. To enable them to be adjusted by a knob, spindle potentiometers (or 'pots') are usually mounted on a panel. The 'preset' type is generally soldered in place in a circuit and then adjusted just once before being left alone. These preset variable resistors are set using a small screwdriver or special adjusting tool. Some preset variable resistors can be adjusted with great precision, and are classed as 10-turn or 20-turn presets depending on how many turns of the adjusting screw are required to move the wiper from one end of the resistance track to the other.

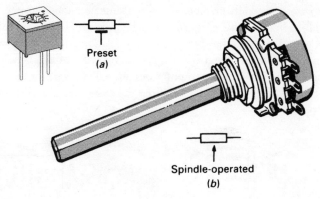

Preset
(a)

Spindle-operated
(b)

Circuit symbols

Figure 5.5 Types of variable resistor: (a) preset type, and (b) spindle-operated.

Insight

At this point we can readily accept resistance as a property
that impedes the flow of electricity through a circuit.
However, it is better to think of resistance as a means of
controlling voltage in a circuit, which is the main point of
this chapter.

5.4 Values and coding of resistors

Most fixed-value resistors are marked with coloured bands so that
their values can be read easily. Four coloured bands are used as
shown in Figure 5.6.

Band 1 gives the first digit of the value.

Band 2 gives the second digit of the value.

Band 3 gives the number of zeros that follow the first two digits.

Band 4 gives the 'tolerance' of the value worked out from the first
three bands.

For example, suppose the bands are coloured as follows:

Band	1	2	3	4
Colour	yellow	violet	red	silver
Value	4	7	00	10%

This resistor has a value of 4700 Ω to within 10 per cent more or less. That is, its value is 4.7 kΩ ± 10%. If the value of this resistor were measured accurately, its resistance should not be more than 4.7 kΩ ± 0.47 kΩ, or less than 4.7 kΩ − 0.47 kΩ, i.e. between 5.17 kΩ and 4.23 kΩ. For most circuit designs, it is unnecessary to use resistors with tolerance less than 5 per cent.

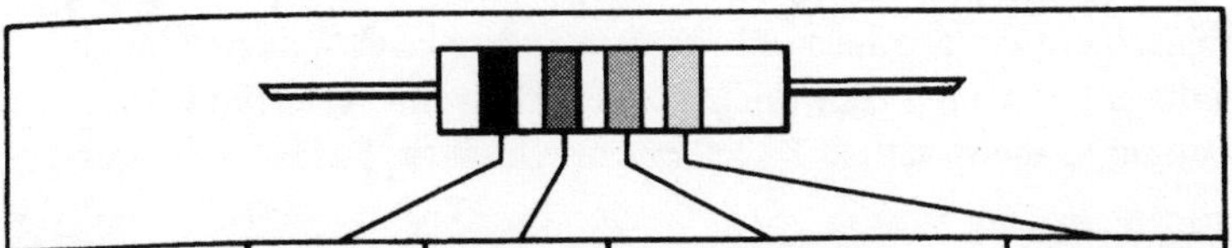

Colour	Band 1	Band 2	Band 3	Band 4
Black	0	0	None	
Brown	1	1	0	1%
Red	2	2	00	2%
Orange	3	3	000	3%
Yellow	4	4	0 000	4%
Green	5	5	00 000	–
Blue	6	6	000 000	–
Violet	7	7	0 000 000	–
Grey	8	8	–	–
White	9	9	–	–
Gold	–	–	0·1	5%
Silver	–	–	0·01	10%
No colour	–	–	–	20%

Figure 5.6 The resistor colour code.

Instead of using a colour code, some manufacturers are marking the values of resistors using the British Standards BS1852 code. This code is often used to mark resistor values on circuit diagrams, as in this book. The BS1852 code consists of letters and numbers as the following examples show.

BS1852 code	Resistance
6K8M	6.8 kΩ± 20%
R47K	0.47 Ω± 10%
5R6J	5.6 Ω± 5%
47KG	47 kΩ± 2%
2M2F	2.2 MΩ± 1%

Note that in the BS1852 code, instead of the decimal point a letter, e.g. 'K', is used to indicate the multiplying factor. Thus in the code 6K8M, the K indicates that the resistor has a value of $6.8 \times 1000 \ \Omega = 6.8 \ k\Omega$.

To avoid making an impossibly large number of resistor values, manufacturers produce only certain values known as preferred values. For 10 per cent tolerance resistors, the preferred values belong to the so-called E12 series in which the twelve values are:

10, 12, 15, 18, 22, 27, 33, 39, 47, 56, 68, 82.

For 5 per cent tolerance resistors, the E24 series includes an additional twelve values as follows:

11, 13, 16, 20, 24, 30, 36, 43, 51, 62, 75, 91.

5.5 Power ratings of resistors

Heat is produced (or dissipated) within a resistor when current flows through it, and this is unavoidable. This heat is generally a nuisance in electronic circuits, but it is put to good use in electric soldering irons, incandescent filament lamps, fire bars and the like. The heat represents the conversion of electrical energy into heat energy. Now like all forms of energy, heat energy is measured in joules. The rate at which heat is produced within the resistor is measured in joules per second and is equal to the electrical power produced in the resistor measured in watts. Thus,

$$\text{power (watts)} = \text{rate energy is produced (joules per second)}$$
$$\text{or } W = J/s \text{ or } Js^{-1}$$

The unit of electrical power, the watt, is used to estimate the power rating of all types of devices, from soldering irons (e.g. 20 W) to power station generators (e.g. 10 MW), from solar panels on spacecraft (e.g. 2 kW) to resistors (e.g. 250 mW). For a resistor, it is a simple matter of calculating the electrical power dissipated within it as heat. It is given by the equation:

$$\text{watts} = \text{volts} \times \text{amps}$$
$$\text{or } W = V \times I$$

In this equation, V is the p.d. across the resistor and I is the current flowing through the resistor. So, suppose a current of 10 mA flows through a resistor when the p.d. across it is 5 V. The heat generated within the resistor is then

$$5 \text{ V} \times (10/1000) \text{ A} = 0.05 \text{ W or } 50 \text{ mW}$$

Note that it is usual to state the electrical power generated within resistors, and other components, in milliwatts (mW) since it is usually less than 1 W. Now to save you the trouble of measuring the current flowing through the resistor, you can use an alternative form of the power equation. You will remember that $V = I \times R$ (Section 3.3). So $I = V/R$, and the power equation becomes

$$W = V \times I = V \times V/R = V^2/R$$

These two quantities, V and R, are readily obtained; R from the colour code on the resistor, or from an ohmmeter, and V from a voltmeter. If you do happen to know the value of the current flowing through the resistor, then use the following alternative form of the power equation

$$W = V \times I = I \times R \times I = I^2 \times R$$

Resistors, both fixed-value and variable, are rated according to the maximum allowable heat generated in them. Exceed this rating, and the resistor is likely to be damaged by its own self-heating. And the same precaution applies to many other electronic devices, such as transistors and diodes that have a maximum safe power rating. General-purpose resistors are rated at 1/4 W (250 mW), 1/2 W (500 mW), 1 W, and 2 W. For most of the applications discussed in this book, 250 mW carbon-film or metal-film resistors are suitable, though there is no harm in using resistors of higher rating.

Suppose a resistor of value 1 kΩ is used in circuit where the p.d. across it is 16 V. The power dissipated in the resistor is

$$W = V^2/R = (16)^2/1 \text{ k}\Omega = 256/10^3 = 256 \text{ mW}$$

Thus a 250 mW resistor is just adequate and you would expect it to be warm to the touch. A 500 mW type would have more than adequate capacity to dissipate this amount of heat.

5.6 Special types of resistor

Some resistors change their resistance in response to the change in some property. Two of the most useful devices of this type are shown in Figure 5.7. The resistance of the light-dependent resistor (LDR) changes with the amount of light falling on it. The resistance of the thermistor changes with its temperature. The graphs show how an ohmmeter records the change in resistance of these devices. The resistance of the LDR increases greatly from daylight to darkness, while the resistance of the thermistor would increase by a much smaller amount if its temperature fell from, say, 100°C to 0°C. Both the LDR and the thermistor are based on semiconductors. The LDR uses a material such as cadmium sulphide, and the thermistor a mixture of different semiconductors. While most LDRs are like the one shown in Figure 5.7a, thermistors may be disc-shaped (as shown in Figure 5.7b) or rod-shaped. For maximum sensitivity, a 'glass bead thermistor' uses a small piece of semiconductor encapsulated in a glass envelope. This is ideal for temperature measurement since it responds quickly to temperature changes.

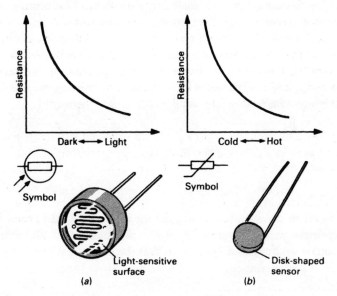

Figure 5.7 Two special types of resistor: (a) the light-dependent resistor, and (b) the thermistor.

Insight

RoHS, which stands for Restriction of Hazardous Substances, is the directive which restricts the use of six hazardous substances in electrical and electronic equipment. The directive was adopted by the European Union in February 2003 and enforced on 1st July 2006. Besides cadmium the other substances are: lead, mercury, hexavalent chromium, polybrominated biphenyls, and polybrominated diphenyl ether. Thus, when you buy components and equipment labelled RoHS from a supplier, you can be sure that the product meets the directive's criteria. For example, for cadmium, used in the LDR, this is 100 parts per million by weight of homogenous material. For further information see: http://en.wikipedia.org/wiki/Restriction_of_Hazardous_Substances_Directive#Health_benefits

Light-dependent resistors are used in photographic light meters, security alarms and automatic street light controllers. Thermistors are used in control systems such as thermostats, in fire alarms and in thermometers.

The LDR and thermistor are generally used as one of the resistors in a potential divider as shown in Figure 5.8. When their resistance changes, there is a change of p.d. across them. Thus, if we use the following relationship (Section 5.1)

$$V_{out} = V_{in} \times \frac{R_2}{R_1 + R_2}$$

In this equation, R_2 is the resistance of the LDR or thermistor. It is easy to see what happens when the resistance of the LDR or thermistor changes. Figure 5.8a shows that when the LDR is in shade, its resistance, R_2, is high so that the p.d. across it (i.e. V_{out}) is high. If it is in sunlight, its resistance falls, so V_{out} falls.

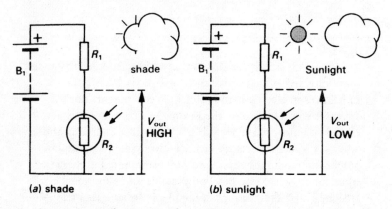

Figure 5.8 Using the LDR in a potential divider.

Similarly, if the thermistor replaces the LDR in this potential divider, high temperature makes the resistance, R_2, of the thermistor low and so V_{out} is low. A low temperature increases the thermistor's resistance so that V_{out} is high. Note that there is a smooth change of output voltage from the potential divider with change of light intensity or temperature. Thus, the potential divider is an analogue device, since V_{out} changes smoothly with change of temperature or light intensity. The potential divider in which one resistor is an LDR or a thermistor is a very useful arrangement in control and instrumentation circuits, usually forming part of a Wheatstone bridge circuit.

5.7 The Wheatstone bridge

In this arrangement two potential dividers are connected in parallel across the same power supply, E. Thus the voltage at X, i.e. the p.d. across R_2, is given by

$$\frac{E \times R_2}{R_1 + R_2} = \frac{E}{1 + R_1/R_2}$$

In addition, the voltage at Y, i.e. the p.d. across R_3, is given by

$$\frac{E \times R_4}{R_3 + R_4} = \frac{E}{1 + R_3/R_4}$$

Thus, if

$$\frac{R_1}{R_2} = \frac{R_3}{R_4}$$

the voltage at X equals the voltage at Y. There is then no potential difference between the points X and Y, and a voltmeter connected between these two points would not show a deflection. The Wheatstone bridge is then 'balanced'. Charles Wheatstone used the bridge to make accurate measurements of resistance, since if three of the resistor values in the above relationship are known, the fourth value can be found.

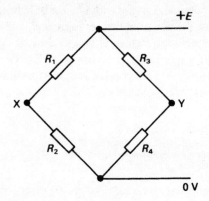

Figure 5.9 The Wheatstone bridge.

The Wheatstone bridge is very useful in electronic measurement and control circuits since a voltmeter placed between the points X and Y is very sensitive to any changes in the resistor values. Suppose the voltmeter is a centre-zero type, (or, better, a digital voltmeter) as shown in Figure 5.10, so that it responds to any difference of voltage between X and Y. In practice, one of the resistors, R_1 say, might have a resistance that varies with temperature, light intensity or pressure. In Figure 5.10, a temperature-sensitive thermistor, Th_1, takes the place of R_1. At a particular temperature, the Wheatstone bridge is first balanced by adjusting the value of VR_1 so that the voltmeter reads 0 V. Now if the temperature rises, there is a decrease in the resistance of the thermistor and the voltmeter will show a reading. The increase of temperature makes the voltage at point X rise above that at point Y. The voltmeter records this increase of voltage. A cooling of the thermistor causes a decrease in the reading. Of course, the voltmeter could be used as a simple electronic thermometer if its scale were calibrated in degrees Celsius.

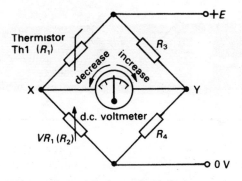

Figure 5.10 Using a thermistor in a Wheatstone bridge for measuring temperature.

The above equation could be used to work out the resistance of the thermistor if the values of VR_1, R_3 and R_4 were known. Thus, the above equation can be written

$$R_1 \text{ (resistance of thermistor)} = \frac{R_3 \times R_2}{R_4}$$

where R_2 is the value of VR_1. In order to detect when the bridge is balanced, the voltmeter should be a sensitive millivoltmeter such as the 0.05 V (50 mV) range on a multimeter though a multimeter would not have a centre-zero reading.

5.8 The strain gauge

Strictly speaking, there is no such phenomenon as an irresistible force meeting an immovable object. A force, however small, always causes an object to yield slightly; it distorts or 'strains' in response to the force. For so-called 'rigid' objects, the strain is so very small that a special type of resistive transducer called a strain gauge is used to measure it. A typical strain gauge is shown in Figure 5.11a. It consists of a metal foil that has a resistance between 60 ohms and 2000 ohms.

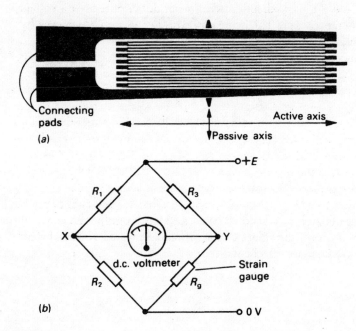

Figure 5.11 (a) The general appearance of a strain gauge, and (b) how it is used with a Wheatstone bridge.

Forming the thin foil is a matter of rolling out an electrically resistive material and then etching parts away (rather like when a printed circuit board is made). The result is a thin flexible resistor in the form of a grid pattern as shown in Figure 5.11a. To use the strain gauge, it is glued to the surface of the object that is undergoing strain. As the object bends, expands or contracts so does the strain gauge. Now if any metal is stretched, its resistance increases; if it is compressed, its resistance decreases. The resistance change is small, perhaps one tenth of an ohm for a 120 Ω strain gauge. However, the Wheatstone bridge can respond to this small change of resistance. Figure 5.11b shows how the strain gauge is connected in a Wheatstone bridge. One voltage divider comprises resistors R_1 and R_2, and the other comprises resistors R_3 and the strain gauge, R_g. A sensitive voltmeter, V, is placed between X and Y.

As you now know, if $R_1 = R_2$ and $R_3 = R_g$, the voltages at points X and Y are equal, and the voltmeter does not register a voltage difference. (This 'balancing of the bridge' is usually achieved by making resistor R_1, R_2, or R_3 a variable resistor.) If the strain gauge is attached to the surface of a material that stretches (called tensile strain), under the action of a force (called tensile stress), its resistance increases slightly. This makes the voltage at Y rise slightly above the voltage at X and the voltmeter shows a deflection one way. Should the strain gauge be compressed slightly (by a compressive force), it contracts slightly (compressive strain) and its resistance falls. This makes the voltage at Y fall slightly below that at X and the voltmeter shows a deflection in the opposite direction. Thus, this simple instrumentation circuit shows which way the object to which the strain gauge is attached is bending. In practice, special instrumentation circuits amplify the small change in voltage between points X and Y. An integrated circuit called an operational amplifier is generally used in these circuits as described in Chapters 13 and 14.

TEST YOURSELF

1 The letters e.m.f., mean...

   extra mighty flow, electrical meeting force, or electromotive
   force?

2 What can you say about the current strength at different
   points in a series circuit?

3 A resistor has coloured bands on it in the order orange, white,
   red and silver. What is its value and tolerance?

4 What is the resistance of a resistor if a voltage of 4.5 V across
   its ends causes a current of 1.5 mA to flow through it?

5 A resistor of value 4.7 kΩ connected in series with an
   unknown resistor provides a total resistance of 10.3 kΩ.
   What is the value of the unknown resistor?

6 Is the following statement true or false?

   A Wheatstone bridge is said to be balanced when no current
   flows through the meter shown in Figure 5.11(b).

7 In the circuit for question 6, R_1 = 100 Ω, R_2 = 680 Ω, and
   R_3 = 200 Ω. The value of R_g that makes the bridge balance
   is ... ohms.
   680, 1500, 1360 or 1280?

8 Describe how the resistance of a light-dependent resistor
   changes with light intensity.

9 Describe the use of a strain gauge.

10 State three applications in which a thermistor might be used.

6

....

Capacitors, timers and oscillators

In this chapter you will learn:
- *the general structure and function of a capacitor*
- *about different types of capacitor*
- *to calculate the resultant value of capacitors connected in series and in parallel*
- *to define the time constant of a capacitor–resistor combination*
- *how to sketch voltage–time graphs for the charge and discharge of a capacitor*
- *the meaning and purpose of a monostable and an astable*
- *to design simple timers and oscillators using capacitors and integrated circuits.*

6.1 What timers and oscillators do

Figure 6.1 shows the functions of two black boxes. First the timer: when this electronic device receives a 'trigger' signal at its input, the voltage at its output rises sharply, becoming HIGH. The output voltage remains HIGH for a time delay of T seconds, and then falls to 0 V again, the output voltage is then LOW. After the output voltage has fallen to 0 V, the timer needs another trigger signal at its input to repeat the time delay. The time delay is determined by the values of components within the timer black box. Timers are used a lot in electronic systems. Toasters, security lights, washing machines, digital clocks and watches, cameras and industrial processes use timers to ensure that a particular operation takes place for prearranged time periods.

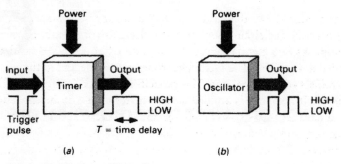

Figure 6.1 The function of (a) a timer, and (b) an oscillator.

Figure 6.1b shows the function of one type of oscillator. When this black box is switched on, its output voltage goes HIGH/LOW continually. The waveform of the signals produced by this oscillator is generally known as a rectangular wave. It is a square wave if the times for which the output voltage is HIGH and LOW are equal. Oscillators that produce these waveforms are used in alarm systems for flashing a lamp on and off, or for sounding an audio alarm from a loudspeaker. They are also used in electronic musical instruments, digital clocks and watches and in computers. The function of timers and oscillators is largely determined by the properties of an electronic component called a capacitor.

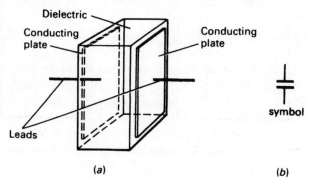

Figure 6.2(a) The basic structure of a capacitor, and (b) its circuit symbol.

6.2 The way a capacitor works

Figure 6.2a shows the basic structure of a capacitor, and
Figure 6.2b the circuit symbol that reflects this structure. It
comprises two metal electrodes separated by an electrical insulator
called a dielectric. The metal electrodes are connected to the
terminals of the capacitor. The capacitor is able to store electric
charge. If a voltage is applied across the terminals of the capacitor
by a battery, B_1, as shown in Figure 6.3a, there is a short flow of
electrons in the external circuit from one electrode to the other.
Thus, one electrode becomes negatively charged and the other
positively charged. The p.d. across the terminals is then equal to
the e.m.f., E, of the battery and the capacitor is said to be charged.
That is, the excess of electrons on one electrode, and the deficiency
of an equal number of electrons on the other electrode, represent
a store of charge. If the battery is removed, these charges remain
in place since they are separated by the dielectric which is an
insulator. Now if the electrodes of the capacitor are joined together
by a conductor as shown in Figure 6.3b, electrons flow in the
reverse direction through the conductor until the p.d. across the
capacitor falls to zero as the charges are neutralized.

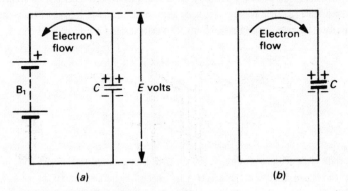

Figure 6.3 The (a) charging, and (b) discharging of a capacitor.

6.3 Units of capacitance

The unit of capacitance is the farad (symbol F). The farad is defined as follows: it is the capacitance of a capacitor that stores a charge of 1 coulomb when it has a p.d. of 1 V across its terminals. In general, if a charge of Q coulombs is given to a capacitor of C farads and the resulting rise in p.d. is V volts, then

$$Q = C \times V$$

The farad happens to be too large a unit to express the values of capacitors generally used in electronics, and it is necessary to use fractions of a farad as follows:

Fraction	Abbreviation
microfarad	µF (10^{-6} F)
nanofarad	nF (10^{-9} F)
picofarad	pF (10^{-12} F)

It is useful to remember that 1000 pF = 1 nF, and 1000 nF = 1 µF. Capacitors in common use range from values as small as, say, 5 pF to as large as 10 000 µF. Suppose a capacitor of value 100 µF is used in a circuit where the p.d. across it is 15 V. What is the charge stored by the capacitor? Using the equation $Q = C \times V$ above:

$$\begin{aligned} Q &= 100 \times 10^{-6}\,\text{F} \times 15\ \text{V} \\ &= 1.5 \times 10^{-3}\text{coulombs} \\ &= 1.5\ \text{millicoulombs} \\ &= 1.5\ \text{mC} \end{aligned}$$

Note that this is a very small charge but it does represent the movement from one plate to the other of the capacitor of a very large number of electrons: 9.375 thousand million million to be precise! In electronics, we are not very concerned with the precise amount of charge a capacitor stores for a given voltage. We are more interested in how the charge storage properties of a capacitor are used in the design of electronic timers and oscillators.

Note: The symbol, F, will be omitted from capacitor values in all of the circuits in this book. Instead capacitor values will be indicated by the multiplier 'p' when referring to values of 10^{-12} F, by the multiplier 'n' when referring to values of 10^{-9} F, and the multiplier 'μ' when referring to values of 10^{-6} F. However, for clarity in the text, the symbol F will be included.

Insight

The farad is named in honour of the nineteenth-century English scientist Sir Michael Faraday, to whom we owe so much for the electrified world of today. He was born on 22 September 1791 in south London, and his family was not well off, and so he received only a basic formal education. However, he educated himself by reading books on a wide range of scientific subjects. In 1821 he published his work on electromagnetic rotation (the principle behind the electric motor). In 1826, he founded the Royal Institution's Friday Evening Discourses and in the same year the Christmas Lectures, both of which continue to this day. He himself gave many lectures, establishing his reputation as the outstanding scientific lecturer of his time.

6.4 Types of capacitor

The value of a capacitor is usually printed on it, or it is marked with a set of coloured bands – see below. Also marked on the capacitor is its maximum safe working voltage (e.g. 100 V). If this voltage is exceeded, a surge of current through the insulating dielectric of the capacitor would certainly damage it. There are many different types of capacitor according to the type of dielectric used in their construction. Figure 6.4 shows five different types of capacitor in general use.

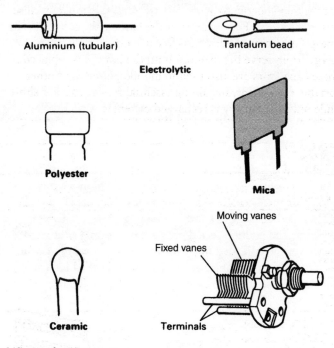

Figure 6.4 Five types of capacitor.

Electrolytic capacitors generally have values in the range 1 µF to 50 000 µF with working voltages up to 100 V, though 450 V ones are available for thermionic valves circuits. They generally have a 'Swiss roll' construction in which the dielectric is a very thin layer of metal oxide between electrodes of aluminium or tantalum foil. The foil is rolled up to obtain a larger area of electrodes in a small volume. A tantalum capacitor offers a high capacitance in a small volume, and some of them are colour-coded as shown in Figure 6.5. Electrolytic capacitors are not suitable for frequencies above about 10 kHz and are commonly used in timing systems and as smoothing capacitors in power supplies (Chapter 7).

Insight

The largest value capacitors are electrolytic and they can store enough energy to act as a battery backup where to lose power is critical. For example, if the battery should fail in the laptop I am using to write this, a capacitor provides enough energy to preserve the memory until the battery is replaced. These capacitors are also used in mobile phones and other portable equipment, sustaining essential functions for a short while until the battery is recharged or replaced.

CAPACITOR COLOUR CODE

	TANTALUM microfarads (μ)				SERIES C280 picofarads (p)				
Band (ring) Colour	1st band	2nd band	Spot (Multiplier)	3rd band	1st band	2nd band	3rd band (Multiplier)	4th band	5th band
BLACK	–	0	1	10V	–	0	1	20%	–
BROWN	1	1	10	–	1 1	1	10	–	100V
RED	2	2	100	–	2	2	100	–	250V
ORANGE	3	3	–	–	3	3	1000	–	–
YELLOW	4	4	–	6·3V	4	4	10 000	–	400V
GREEN	5	5	–	16V	5	5	100 000	5%	–
BLUE	6	6	–	20V	6	6	1 000 000	–	–
VIOLET	7	7	–	–	7	7	0·01	–	–
GREY	8	8	0·01	25V	8	8	0·01	–	–
WHITE	9	9	0·001	30V	9	9	–	10%	–
PINK	–	–	–	35V	–	–	–	–	–

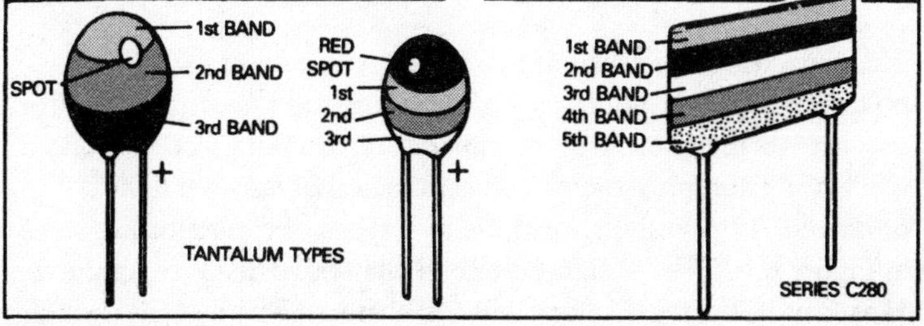

Figure 6.5 Colour coding of tantalum electrolytic and C280 polyester capacitors.

Polyester capacitors are examples of plastic-film capacitors. Polypropylene, polycarbonate and polystyrene capacitors are also types of plastic-film capacitor. The plastic, with the exception of polystyrene that has a low melting point, has a metallic film deposited on it by a vacuum evaporation process – these are called metallized capacitors. Metallized polyester film capacitors can have values up to 10 μF, but other plastic-film capacitors have lower values. Working voltages of polyester capacitors can be as high as 400 V, and of polycarbonate capacitors, 1000 V. The Mullard C280 series of polyester capacitors were colour-coded as shown in Figure 6.5 but this way of marking the values is now obsolete.

Mica capacitors are rather more expensive than plastic-film capacitors and they are made by depositing a thin layer of silver on each side of a thin sheet of mica. Like ceramic capacitors consisting of a silver-plated ceramic tube or disc, and polystyrene capacitors, they are excellent for use at high frequency. Mica capacitors have values in the range 1 pF to 10 nF, have excellent stability and are accurate to +1% of the marked value.

Variable capacitors have low maximum values, e.g. 500 pF, and their construction involves one set of metal plates moving relative to a fixed set of metal plates. The plates are separated by air or a plastic sheet that acts as the dielectric. Variable capacitors are often used in radio receivers for tuning in to different stations.

6.5 Capacitors in parallel and series

The capacitors described above fall into two main categories: polarized and non-polarized. Aluminium and tantalum capacitors, that are normally electrolytic capacitors, are polarized (see above), while polyester, polystyrene and ceramic capacitors are non-polarized. The symbols for these two types of capacitor are shown in Figure 6.6. The '+' sign on one of the terminals of the polarized capacitor indicates that this capacitor must be connected the right way round in a d.c. circuit, this leading to the positive supply or signal voltage. Note that the two parallel lines of the symbol separated by a space (filled with a dielectric) are indicative of the construction of a capacitor.

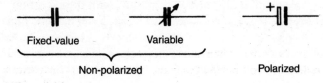

Figure 6.6 Three symbols for capacitors.

If the area of each plate of a capacitor is doubled, their separation remaining the same, the capacitance of the capacitor is doubled since the area is able to store twice the charge. In addition, if the separation of the plates is halved, their area remaining the same, the capacitance is also doubled. These relationships give us a clue to the capacitance of two capacitors connected in parallel as shown in Figure 6.7a. The total area of the plates is effectively the sum of the two areas, so the total capacitance C of the combination is found by adding their values together.

$$C = C_1 + C_2 \text{(parallel)}$$

Thus, if two capacitors of 10 µF and 50 µF are connected in parallel, the combined capacitance is 60 µF. However, if two capacitors are connected in series as shown in Figure 6.7b, the formula for finding their total capacitance is

$$C = \frac{C_1 \times C_2}{C_1 + C_2} \text{(series)}$$

The combined capacitance is equal to the product of their individual values divided by the sum of their individual values. Accordingly, if the above two capacitors of 10 µF and 50 µF are connected in series, the total capacitance of the combination is given by

$$C = \frac{10 \times 50}{10 + 50} = \frac{500}{60} = 8.33 \ \mu F$$

The combined capacitance of three capacitors connected in series is given by the equation

$$\frac{1}{C} = \frac{1}{C_1} + \frac{1}{C_2} + \frac{1}{C_3}$$

and so on. Note that when two or more capacitors are connected in parallel, their combined capacitance is more than the capacitance of the largest value. When they are connected in series, their combined capacitance is less than the smallest value capacitor. You should compare these equations with the equations for series and parallel resistor combinations in Chapter 5.

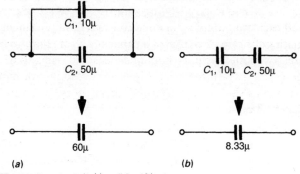

Figure 6.7 Two capacitors connected in (a) parallel, and (b) series.

6.6 Charging and discharging capacitors

Many timers and oscillators are based on the simple circuit shown in Figure 6.8. Here a capacitor, C_1, and a resistor, R_1, are connected in series with a battery of e.m.f., E. On closing the switch, SW_1, the p.d. across the capacitor rises as shown by the graph. The p.d. rises relatively fast at first and then more slowly as the p.d. approaches the e.m.f., E, of the battery. The delay in the charging (and discharging) of a capacitor is the key to understanding the use of capacitors in the design of timers and oscillators. The time taken for the p.d. across the capacitor to rise to two-thirds of E is known as the time constant of the capacitor/resistor circuit, or RC circuit. The time constant, T, is dependent on the values of both R_1 and C_1 and is given by the simple equation

$$T = R_1 \times C_1$$

This equation gives the time constant in seconds if C_1 has a value in farads and R_1 has a value in ohms. For example, suppose the two values are C_1 = 1000 µF and R_1 = 10 kΩ. Thus C_1 = 1000 × 10^{-6} F and R_1 = 10 × 10^3 = 10^4 Ω. Therefore, T = 10^{-3} F × 10^4 Ω = 10 s. This means that if E = 9 V, it takes 10 seconds for the p.d. across the capacitor to rise from 0 V to two-thirds of E (6 V). In practical timer and oscillator circuits, the time constant is a convenient way of estimating the rate of charge of a capacitor. Note that, to be completely accurate, the p.d. should be measured rising to 63% of E. However, 'two-thirds' (67%) is accurate enough for most practical circuits since many electrolytic capacitors have a tolerance of +50% or more.

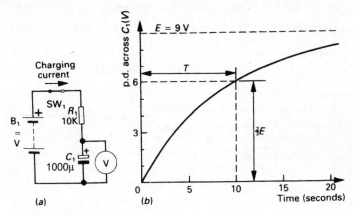

Figure 6.8 The time constant of a RC combination.

Just to complete our examination of the RC circuit, suppose the capacitor is discharged once it has been charged. Figure 6.9 shows what happens if the battery is isolated and the capacitor is allowed to discharge through resistor R_1. The graph shows that the p.d. across the capacitor falls fast at first and then more slowly as the voltage across it approaches 0 V. Incidentally, the graphs for the charge and discharge of a capacitor are known, mathematically, as exponential curves. We do not need to look at the mathematical equation for these graphs and how they help to define the time constant, but they do have an interesting property. This is shown in Figure 6.10 for the charging curve.

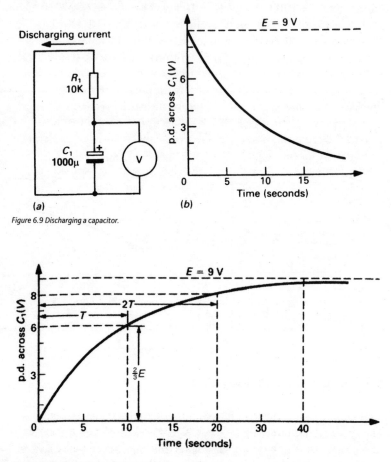

Figure 6.9 Discharging a capacitor.

Figure 6.10 The effects of successive time constants.

Suppose we carried on timing the charging of the capacitor after it had reached two-thirds of E (6 V). We should find that after another time constant of 10 s, the p.d. across the capacitor would have risen by two-thirds of the remaining p.d. Since the remaining p.d. is 3 V (9 V – 6 V), the p.d. would rise by a further 2 V to 8 V. After the next time constant of 10 s, the p.d. across the capacitor would have risen to two-thirds of 1 V (9 V – 8 V), i.e. by 0.67 V to 8.67 V. And so on. You can see that after three time constants (after 30 s in the example), the capacitor is almost fully charged, the p.d. across it having risen to 8.67 V, only 0.33 V short of its final p.d. Theoretically, the capacitor can never become fully charged (a characteristic of exponential curves), but after five or so time constants we can consider it to be fully charged.

Insight

A good example of exponential decay rather than growth is the decay of a radioactive substance such as radium. Radioactive substances decay according to a 'half-life'. This is the time it takes for half the substance to decay. For example, if the half-life is 10 minutes, half the substance decays in 10 minutes. In the next 10 minutes, half of the remaining substance decays and so on. This implies that at any time in the future there is always some undecayed substance remaining, small though it is. It's like taking a journey from A to B. On the first day you cover half the distance; on the second day half of the remainder and so on. Will you ever reach your destination at B? Can you think of other examples of exponential growth or decay?

6.7 The 555 timer monostable

Figure 6.11 shows the systems diagram of an electronic timer. The timer is 'triggered' by an input signal and produces an output signal that lasts for a time T seconds. Now in order to make this function possible we need to connect together three black boxes as shown in Figure 6.11:

(a) *black box 1 is the RC combination discussed above;*
(b) *black box 2 is a device that detects when the p.d. across the capacitor has reached a certain value;*
(c) *black box 3 is an output circuit that indicates that timing is in progress.*

Now black box 2 must be a device that:

(a) *makes the output signal go HIGH when it receives a trigger signal, and*
(b) *returns the output signal to LOW when it detects that the p.d. across the capacitor in the RC combination has risen by a certain amount.*

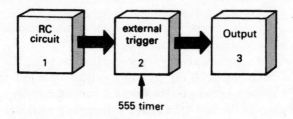

Figure 6.11 The electronic system of a timer.

Nowadays, a circuit designer would choose an integrated circuit (IC) to carry out the function of black box 2. And the most popular of the ICs available for this job is the device designated by manufacturers as the '555 timer' or 'triple-5 timer'. It is shown in Figure 6.12, and comprises a small black plastic package having eight terminal pins for connecting it into a circuit. A silicon chip (Chapter 13) about 2 mm by 2 mm in area is inside the package.

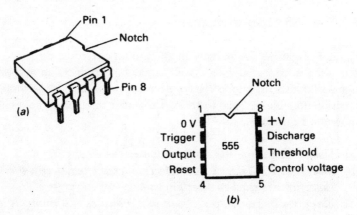

Figure 6.12 The integrated circuit 555 timer: (a) what it looks like, and (b) the identity of its pins.

Now we do not need to know anything about the workings of this chip to use it as black box 2 in the timer system. Figure 6.13 shows the circuit equivalent of the three interconnected black boxes. The main features of this circuit are as follows.

On closing switch SW_1 to switch on the power supply, the output signal is LOW, and the lamp L_1 is switched off. At this point, the capacitor C_1 is discharged (by a transistor on the 555 timer chip). Momentary pressing of the push-to-make, release-to-break switch, SW_2, triggers the timer. This opens the internal switch of the 555 timer and simultaneously makes the output of the 555 timer go HIGH. The lamp switches on and the capacitor C_1 begins to charge through resistor R_1. The lamp remains lit until the p.d. across the capacitor reaches two-thirds (exactly) of the e.m.f. of the supply. When the voltage across the capacitor reaches 6 V (two-thirds of 9 V), this is sensed by terminal pin 6 of the 555. At this instant, the internal switch on the 555 closes, C_1 is instantly discharged via pin 7, and the output of the 555 goes LOW. The lamp switches off and the circuit now waits for another trigger signal.

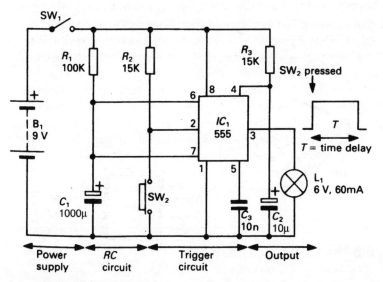

Figure 6.13 A timer based on the 555 IC.

The time that the lamp remains lit is given by the equation $T = 1.1 \times C_1 \times R_1$. So, using the values in Figure 6.13,

$$T = 1.1 \times (1000 \times 10^{-6} \text{ F}) \times (100 \times 10^3 \ \Omega)$$
$$= 110 \text{ s}$$

This calculation assumes that the 1000 µF capacitor actually has the value marked on it, but we know that electrolytic capacitors have big tolerances. A second problem with large-value electrolytic capacitors is that some of the charging current leaks through the capacitor, from one plate to the other. These practical problems generally make the time delay longer than calculated. Therefore, it is often best to replace R_1 by a variable resistor of value 100 kΩ and adjust the time delay to precisely 100 s. However, once set, the 555 timer will deliver the same delay every time it is triggered, even though the e.m.f. of the power supply may fall with time.

If you want the timer to switch on more power, a relay (Chapter 3) should replace the lamp. The contacts on the relay should be rated to switch the current required by the high-power load. Incidentally, the function of resistor R_2 is to make sure that the trigger input (pin 2) is held HIGH before and after switch SW_1 is operated. And components R_3 and C_2, connected to pin 4, ensure that the timer does not begin to operate when the power supply is switched on. In fact, pin 4 could be used to reset the output to LOW and turn the lamp off, at any time during a delay. Capacitor C_3 ensures reliable operation of the timer in an electrically noisy environment.

Finally, note that this timer is properly called a monostable (strictly monostable multivibrator), the word meaning 'one stable state', i.e. the output waveform is stable only when it is LOW. The HIGH state during the time delay is a temporary and unstable state of the timer.

6.8 The 555 timer astable

As explained in Section 6.1, the waveform produced by a rectangular wave oscillator comprises a continuous series of HIGH and LOW signals. Figure 6.14 shows the systems diagram for a simple rectangular wave oscillator. It is based on three black boxes:

(a) *black box 1 is the RC combination described above;*
(b) *black box 2 is a device that detects when the p.d. across the capacitor has reached a charged and a discharged value;*
(c) *black box 3 is an output circuit that makes use of the HIGH/LOW output signals.*

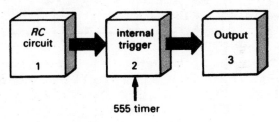

Figure 6.14 The electronic system of an oscillator.

The 555 timer IC can be made to operate as an oscillator as well as a timer. Figure 6.15 shows a circuit that performs the function of the three interconnected black boxes. The circuit is slightly more complicated than the timer circuit of Figure 6.13. Note that a resistor R_2 is now connected between pins 6 and 7 of the 555 timer. In addition, pin 4, the reset pin, is connected directly to the power supply. The 555 timer takes the part of black box 2 in the systems diagram of Figure 6.14. The circuit behaves as follows.

On switching on the power supply, the p.d. across the capacitor C_1 is 0 V, and the internal transistor switch is open. At this point, the output waveform is HIGH and the lamp L_1 is lit. The p.d. across C_1 immediately begins to rise as it charges through resistors R_1 and R_2. At the instant that the p.d. across C_1 reaches exactly two-thirds of the supply e.m.f., in this case 6 V, the internal transistor switch closes and C_1 discharges via resistor R_2 and pin 7 of the 555 timer. Simultaneously, the output waveform goes LOW, and L_1 goes out. The output waveform remains LOW as the p.d. across C_1 falls. Pin 6 of the 555 timer senses when the p.d. has fallen to one-third of the supply e.m.f., in this case 3 V. At this instant, the internal transistor switch opens, the output waveform goes HIGH once again, L_1 lights, and the capacitor charges once again through resistors R_1 and R_2. The cycle repeats, the capacitor alternately charging and discharging between 3 V and 6 V, and the lamp switching on and off as the waveform goes HIGH and LOW, respectively.

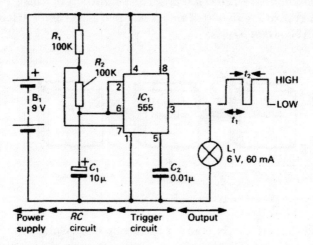

Figure 6.15 An oscillator based on the 555 timer IC.

Note: C_1 charges through R_1 and R_2, and discharges through R_2. This is reflected in the two equations for the HIGH (C_1 charging through R_1 and R_2), and LOW (C_1 discharging through R_2) times of the output waveform from the oscillator. These times are as follows:

$$\text{HIGH time, } t_1 = 0.7 \times (R_1 + R_2) \times C_1$$
$$\text{LOW time, } t_2 = 0.7 \times R_2 \times C_1$$

Thus, using the values of C_1, R_1 and R_2 in Figure 6.15, the t_1 and t_2 times are as follows:

$$t_1 = 0.7 \times (200 \times 10^3) \times 10 \times 10^{-6} = 1.4 \text{ s}$$
$$t_2 = 0.7 \times (100 \times 10^3) \times 10 \times 10^{-6} = 0.7 \text{ s}$$

The lamp flashes on for 1.4 s and off for 0.7 s. In practice, these times might be different from the calculated values due to the tolerance of the components used. However, if precise HIGH/ LOW times are required, it is easy to replace R_1 or R_2, or both, by variable resistors. Note that it is impossible with this simple circuit to get the HIGH and LOW times equal, and therefore to produce a waveform that is a square wave. If C_1 has a value of 100 µF, R_1 and R_2 remaining the same, the HIGH and LOW times each become ten times longer, i.e. 14 s and 7 s. If the lamp is replaced by a relay, the relay contacts could be used to operate higher-power lamps, motors, etc. (If you are puzzled about the factor '0.7' in the above equations, do not be: in the proof of these equations, a factor '0.693' appears and I have approximated this to '0.7'.)

Note that this oscillator is properly called an astable (strictly astable multivibrator), the word 'astable' meaning 'no stable state', i.e. the output is not stable in either of the two output states, HIGH or LOW. It oscillates from one state to the other.

Insight

There are three forms of multivibrator: monostable, astable and bistable. The first two you have met in this chapter and the bistable (or flip-flop) you will meet in Chapter 10. Generally speaking a multivibrator employs positive feedback to cross-couple two devices so that two distinct states are possible. For example, one device ON and the other device OFF, and in which the states of the two devices can be interchanged either by use of external pulses as in the monostable and bistable, or by internal coupling via a capacitor as in the astable. When the circuit is switched between states, transition times are normally very short compared to the ON and OFF periods. Hence, the output waveforms are essentially rectangular in form.

TEST YOURSELF

1 *How would you show that a capacitor stores electrical energy?*

2 *A 30 μF capacitor and a 60 μF capacitor are connected in series. What is the value of the resultant capacitor?*

3 *A 100 nF capacitor is connected in parallel with a 47,000 pF capacitor. What is the total capacitance in nanofarads and microfarads?*

4 *Calculate the charge stored by a 50 μF capacitor when a potential difference of 20 V is applied across its terminals.*

5 *What is meant by time constant? What is the time constant of a 220 μF capacitor connected in series with a resistor of value 22 kΩ?*

6 *Explain the general shape of the charge and discharge, voltage–time characteristics for a capacitor in series with a resistor.*

7 *Distinguish between an astable and monostable circuit.*

8 *Describe one application for a monostable and one for an astable.*

9 *Design a circuit using a 555 timer that will energize a relay for a period of 10 s at intervals of 3 seconds. What is the name for this circuit?*

10 *Design a general-purpose timer based on a 555 timer that will energize a relay for a period of 4 minutes. What is the name of this circuit?*

7

....

Diodes and rectification

In this chapter you will learn:
- *to describe the appearance and properties of rectifier diodes*
- *how a rectifier diode works based on semiconductor construction*
- *what is meant by forward bias and reverse bias for a diode*
- *to draw a V–I characteristic for a semiconductor diode*
- *to draw half-wave and full-wave rectification circuits that use rectifier diodes*
- *to describe the properties of a Zener diode with reference to its V–I characteristic*
- *about the uses of diodes in a simple a.c. to d.c. power supply*
- *about the properties and use of light-emitting diodes*
- *to explain the use of thyristors and triacs for controlling electrical power.*

7.1 What diodes do

Rectification involves the use of one or more components called diodes or simply rectifiers. They behave as one-way 'valves' to control the current flowing through them. Indeed, the first diodes were thermionic valves (Chapter 1), so called because they allow electrons to pass easily through them in one direction only and not in the opposite direction. The symbol for a diode is shown in Figure 7.1a. Two types of diode are shown in Figure 7.1b, one that allows a maximum current of 1 A to flow through it and the other a maximum of 13 A. Note that the arrow of the diode symbol points in the direction of conventional current flow through it, so electrons flow through the diode in the direction opposite to this arrow. Conventional current flows through a diode from its anode terminal to its cathode terminal. On diodes encased in plastics, the cathode terminal is often marked with a red, black or white band. When the anode of a diode has a more positive voltage than the cathode, the diode is said to be forward biased and current flows easily through it (Figure 7.2a). When the anode is more negative than the cathode, it is said to be **reverse biased** and no current flows through it (Figure 7.2b).

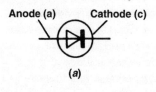

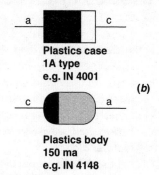

Figure 7.1 Rectifier diodes: (a) general symbol, and (b) the appearance of two types.

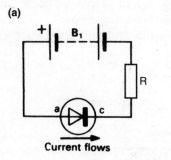

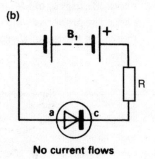

Figure 7.2 (a) A forward-biased diode, and (b) a reverse-biased diode.

7.2 The p–n junction

The rectifying action of a semiconductor diode is due to the electrical effects that occur at the junction between an n-type semiconductor and a p-type semiconductor, a region known as a p–n junction. Figure 7.3a shows that an interesting thing happens when this p–n junction is made: electrons flow across the junction from the n-type material to the p-type material, and holes flow in the opposite direction. The reason for this flow of charge should be clear from what has been said about the properties of p-type and n-type semiconductors (Chapter 2). Electrons move to fill holes and holes move to capture electrons, a process called diffusion.

The movement of electrons and holes across the junction depletes, or reduces, the n-type material of holes and the p-type material of electrons. A narrow depletion layer is created (Figure 7.3b) that is less than one millionth of a metre wide. The resulting positive charge in the n-type material and negative charge in the p-type material produce what is known as a potential barrier, an internal property of the p–n junction that opposes the further diffusion of charge across the junction. Depending on the polarity of an external p.d., the potential barrier is increased or decreased as follows.

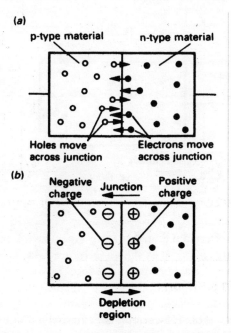

Figure 7.3 (a) Charge flow across an unbiased p–n junction, and (b) the formation of a depletion layer.

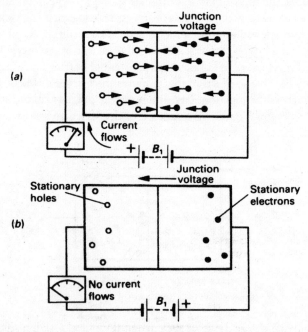

Figure 7.4 (a) A forward-biased p–n junction, and (b) a reverse-biased p–n junction.

Suppose an external p.d. makes the p-type material more positive than the n-type material as shown in Figure 7.4a. The effect is to decrease the potential barrier and reduce the width of the depletion region. If the external p.d. is less than about 0.5 V (for a p–n junction based on silicon), no current flows from the p-type to the n-type material. However, if the external p.d. is increased to about 0.6 V, the depletion region vanishes and current flows across this forward-biased p–n junction. If the external p.d. is applied to the p–n junction as shown in Figure 7.4b, electrons and holes move away from the junction, the depletion region widens and no current flows across this reverse-biased p–n junction. In an actual diode, the anode terminal is connected to the p-type material and the cathode terminal to the n-type material. Thus, the p–n junction acts as a 'one-way valve' for the flow of current through it.

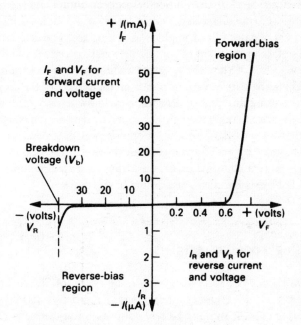

Figure 7.5 The voltage/current characteristic of a silicon junction diode.

If measurements are made of the current flowing through a silicon diode for different values of the p.d. across it, a graph similar to that shown in Figure 7.5 is obtained. The graph is known as the voltage–current, or $V–I$, characteristic of a p–n silicon diode. It is clear from the graph that the current flowing across the p–n junction increases sharply as the external voltage changes from about 0.5 V to 0.7 V. This means that the resistance of the p–n junction falls sharply over this voltage range. In the reverse-bias region, no current flows through the diode until the voltage reaches the reverse breakdown voltage, V_b. This voltage varies from a few volts to a few hundred volts depending on the type of diode used. The diode is usually ruined if V_b is exceeded, but not in the Zener diode which is designed to operate in this region – see above.

Insight

The detection of radio signals requires a diode to rectify the signals (see Section 15.6). Long before germanium and silicon diodes were manufactured, people made simple radios using a crystal of galena (lead sulphide) as a detector. The detection takes place at the point of contact between the crystal and the tip of a piece of wire, the operator finding the best reception by touching the wire at various points on the surface of the crystal. Radios using this kind of detector were known as crystal radios and the wire making the point of contact called a cat's whisker. These radios worked without external power provided that they had a long aerial and a good earth. Even after thermionic valves were used in radios, crystal radios remained popular, and hobbyists find them fascinating to this day.

7.3 Using diodes as rectifiers

An oscilloscope (Chapter 4) is used in the circuit of Figure 7.6 to see how a diode changes, or rectifies, a.c. to d.c. A 50 Hz alternating current supply of a few volts maximum value is connected across diode, D_1, which is also connected in series with resistor, R_1. The oscilloscope monitors the changes of voltage across resistor R_1. With SW_1 closed to short-circuit D_1, the oscilloscope displays an a.c. waveform, as might be expected since it is connected across the a.c. supply. When SW_1 is opened, the trace shows that one half of the a.c. waveform is removed by the diode. This happens because D_1 is alternately forward-biased and reverse-biased by the a.c. supply. The diode allows current to flow through it and R_1 only when terminal T_1 of the supply is positive with respect to terminal T_2. The current through R_1 is a varying d.c. and the circuit is called a **half-wave rectifier**. Note that the half-wave rectifier 'loses' one half of the a.c. waveform.

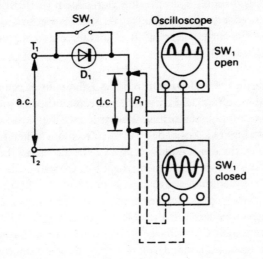

Figure 7.6 Half-wave rectification.

A better rectifier circuit is shown in Figure 7.7; this is known as a full-wave rectifier. The 'lost' half wave now contributes to the d.c. output of the rectifier circuit since current flows through resistor R_1

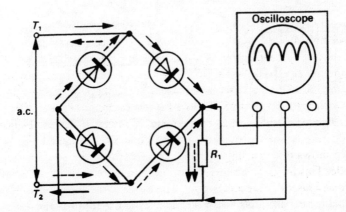

Figure 7.7 Full-wave rectification.

in the same direction for both halves of the a.c. waveform. Thus, when terminal T_1 of the a.c. supply is positive with respect to T_2, current flows through R_1 in the direction shown by the full-line arrows. Moreover, when T_2 is positive with respect to T_1, the direction of the current through R_1 is unaltered and is represented by the dotted-line arrows.

The outputs of the half-wave and full-wave rectifier circuits are a varying d.c. voltage and they need to be smoothed to produce a steady d.c. supply. Smoothing is achieved with a capacitor, C_1 as shown in Figure 7.8 for the half-wave rectifier. When terminal T_1 is positive, the diode conducts and C_1 charges up to the peak voltage of the a.c. supply. During this time, current is also flowing through R_1. When the voltage at T_1 begins to fall during the first half cycle of the supply, current continues to flow through R_1, and C_1 discharges through it. If the time constant, $T = C_1 \times R_1$ (Chapter 6), is long enough, C_1 discharges little during the second half cycle when D_1 is reverse-biased. The effect is a smoothed d.c. supply as shown by the graphs.

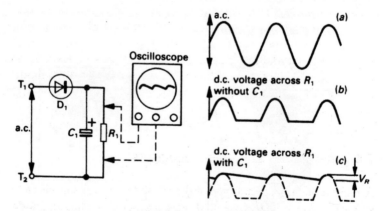

Figure 7.8 Using a smoothing capacitor. The graphs show: (a) unrectified a.c.; (b) half-wave rectified d.c.; (c) smoothed a.c.

The d.c. output voltage across R_1 has a small ripple voltage superimposed on it caused by the small fall in voltage across C_1

during the second half cycle. For good smoothing action, the product $R_1 \times C_1$ must be larger than 1/50 second (0.02 s), which is the time between peak voltages across R_1 for a 50 Hz a.c. supply. For efficient smoothing of the d.c. output, the product $R_1 \times C_1$ ought to be about five times larger than 0.02 s. Thus if $R_1 = 1$ kΩ, $C_1 \times R_1 = 0.1$ s and, since $C_1 = T/R_1$

$$C = \left(0.1/R_1\right) = \left(0.1/10^3\right) = 10^{-4}\,\text{F} = 100\ \mu\text{F}$$

In practical rectifier circuits, smoothing capacitors are of the electrolytic variety (Chapter 6) and have values between about 470 μF and 2200 μF. Perhaps you can see why a capacitor of lower value would suffice to smooth the varying d.c. voltage produced by a full-wave rectifier?

In practical rectifier circuits, it is important to use a diode that will not be damaged by the current flowing through it. This current is called the maximum forward current and it is usually indicated by the symbol I_F. Also note that, since the capacitor discharges very little during the half cycle when the voltage at T_1 changes from its maximum positive to its maximum negative value, the p.d. across the diode reaches almost twice the peak voltage of the a.c. supply. The diode has to be rated to withstand this maximum reverse voltage (V_R) without being damaged. For example, the 1N4001 has an I_F value of 1 A, and a V_R value of 50 V; and the 1N4006 diode also has an I_F value of 1 A, but a V_R value of 800 V.

7.4 Zener diodes as voltage regulators

A zener diode is a special type of diode used in power supplies to stabilize a d.c. voltage. Like all diodes, a zener diode has an anode terminal and a cathode terminal. The characteristic 'knee' in the line at the cathode terminal of the symbol (Figure 7.9a) indicates its special function. The voltage–current characteristic of a typical zener diode shown in Figure 7.10 reflects this 'knee'. The graph shows that the zener diode is operated in the reverse-bias breakdown region where a normal rectifier diode would be damaged. In this region, large current changes through it produce very little voltage change across it, i.e. the slope of the characteristic is very steep. To maintain the zener diode in the reverse-breakdown region, the current through it must not fall below about 5 mA (for this diode).

Figure 7.9 (a) The symbol for a zener diode, and (b) the appearance of one type of zener diode.

Insight

The zener diode was named after Dr Clarence Zener who discovered the breakdown voltage when a zener diode is reverse biased. His early ideas about solid state physics contributed to modern computer circuitry. In particular, in 1934 he wrote a paper explaining the breakdown of electrical insulators that resulted in the 1950s in the zener diode, a voltage regulator that found its uses as a basic component in modern electronics.

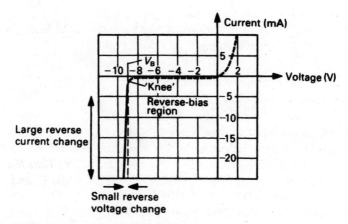

Figure 7.10 The voltage–current characteristic of a zener diode.

The reverse breakdown voltage, V_B, of a zener diode is usually marked on it. Thus if '5V6' is printed on a zener diode, it will stabilize a voltage of 5.6 V. The values of V_B follow a series of preferred values (just as for resistor values, see Chapter 3). These values are 2V7, 3V, 3V3, 3V9, 4V3, 4V7, 5V1, 5V6, 6V2, 6V8, 7V5, 8V2, 9V1, 10V, and so on.

A simple voltage regulator using a zener diode is shown in Figure 7.11. This circuit stabilizes the output voltage at 5.6 V for a wide range of input voltages. There are three main factors to consider in this simple design: the minimum input voltage for which the output voltage remains constant at 5.6 V; the power dissipation of the diode at the maximum input voltage; and the maximum current that can be drawn by the load resistor R_L to maintain voltage stabilization for a specified input voltage. Let us consider each requirement in turn, and assume that the diode has the following ratings: $V_B = 5.6$ V; maximum power dissipation = 500 mW; minimum reverse current = 5 mA. The BZ × 85 series of zener diodes have a power rating of 1.3W.

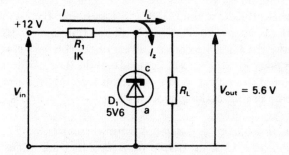

Figure 7.11 A simple voltage-regulated supply.

1 **Minimum input voltage.** *Suppose the load current I_L is zero (so $I = I_z$). First, calculate the p.d. across R_1. If $I_z = 5$ mA and $R_1 = 1$ kΩ, the p.d. across $R = 5$ mA × 1 kΩ = 5 V. Thus the minimum input voltage is the sum of the p.d. across R_1 and the p.d. across D_1, so $V_{in} = 10.6$ V.*

2 **Power dissipation in the diode.** *Suppose the maximum input voltage is 20 V. This input voltage makes the p.d. across R_1 =*

$20 - 5.6 = 14.4$ V. *Thus the current through* $R_1 = 14.4$ V/1kΩ = 14.4 mA = I_z. *The power dissipation, P, in the diode is given by*

$$P = I_z \times V_z = 14.4\ mA \times 5.6\ V = 80.6\ mW,$$

a value well within the maximum rating of 500 mW

3 **Maximum load current.** *Suppose* $V_{in} = 12$ V. *Now p.d. across* $R_1 = 12 - 5.6 = 6.4$ V. *Current through* $R_1 = 6.4$ V/1kΩ = 6.4 mA. *If at least 5 mA of this has to flow through* D_1 *to maintain it in the breakdown region, the maximum load current is 1.4 mA and is very small. Robbing the diode of more current than this would take the operating point of the diode below the 'knee' of its characteristic, and voltage stability would be lost.*

Of course, you could allow more current to be drawn by the load from this simple stabilizer by using either a higher power (and more expensive) zener diode (e.g. the 1.3 W BZY61 series), or by reducing the value of the series resistor R_1 (provided the maximum power rating of the diode is not exceeded). Fortunately, the more elegant solution shown in Figure 7.12 is to hand. This is a practical circuit that provides a nominal 9 V, 2 A stabilized d.c. supply from the a.c. mains. The circuit uses an npn transistor (Chapter 8) to increase the available output current of the voltage stabilizer without placing extra current demands on the diode. The general operation of this power supply can be understood by noting that it is made up of four building blocks.

1 *Building block 1 is a step-down mains transformer* T_1 *that produces a 12 V a.c. output from the 240 V a.c. mains supply.*
2 *Building block 2 is a bridge rectifier,* Br_1, *that contains four silicon rectifier diodes and produces a full-wave rectified d.c. output from the 12 V a.c. supply.*
3 *Building block 3 comprises a smoothing capacitor* C_1 *that produces a smoothed d.c. output of about 14 V. This voltage will vary with variations in the mains voltage.*
4 *Building block 4 is based on a zener diode* D_1 *and npn transistor* Tr_1.

Now the operation of the transistor is described in Chapter 8. Its main purpose here is to provide a higher current than is possible with the simple circuit shown in Figure 7.11. Indeed the output current is increased by the current gain of the transistor. Resistor, R_1, continues to provide the small reverse current for the diode plus that required by the base of the transistor. The zener diode, D_1, holds the voltage at the base terminal of Tr_1 at 10 V. The voltage at the emitter of the transistor is 0.6 V less than this (9.4 V). The amplified current flowing between the collector and emitter terminals of the transistor is the current drawn by the load, and this is limited to 2 A by the fuse. In order to protect the transistor from damage due to the heat generated within it, it is usual to mount it on a heat sink, a chunky piece of copper that has fins on it to help dissipate the heat. The inset pictures show the appearance and pin connections of the bridge rectifier and transistor.

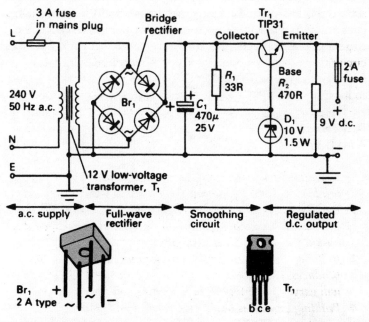

Figure 7.12 A mains-operated d.c. supply.

Nowadays, circuit designers tend to use purpose-designed voltage
regulators that contain the zener diode, associated resistors and
current-amplifying transistors in a single package. Two typical voltage
regulators are shown in Figure 7.13a. These replace the zener diode,
resistor R_1 and transistor in Figure 7.13. A 12 V, 2 A voltage regulator
such as this has internal circuitry that automatically limits the output
current to 2 A maximum (called current limiting), and reduces the
current passing through it if it overheats (called thermal shutdown).

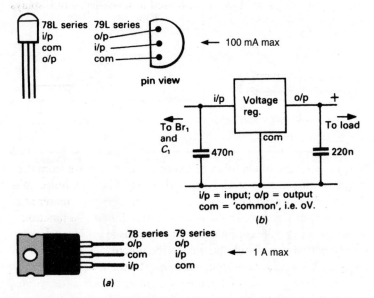

Figure 7.13 (a) Two examples of packaged voltage regulators. (b) The way a voltage regulator is used.

7.5 The light-emitting diode (LED)

By careful selection of p-type and n-type semiconductors, it is possible to make a p–n junction emit light when it is forward biased. This type of diode is called a light-emitting diode (LED), and has generally replaced the fragile, short-life incandescent lamps used as indicator or on/off lamps. In circuits, the LED is designated by a standard diode symbol with two arrows pointing away from the cathode (Figure 7.14b). The life expectancy of the LED is very long – up to 100,000 hours of operation. The shape of a typical LED is shown in Figure 7.14 to indicate light leaving the diode. In one common type of LED, a 'flat' where the leads enter the LED identifies the cathode. Sometimes LEDs have a longer anode terminal than the cathode terminal. It is only necessary for a current of a few milliamperes to pass through the p–n junction to make the LED emit light. Thus, LEDs are ideal as indicators in all types of application, especially for battery-powered devices where a filament lamp would dissipate too much power. Also now available are bright blue-white diodes which are grouped in small clusters to provide an 'everlasting' torchlight. Groups of LEDs are also used in seven-segment displays and bargraph displays as explained in Chapter 11.

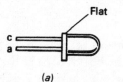

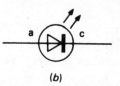

(a) *(b)*

Figure 7.14 (a) Shape, and (b) symbol of a typical LED.

When a p–n junction is forward-biased, electrons move from the n-type to the p-type material where they combine with holes quite close to the junction. Holes also move from the p-type material to combine with electrons in the n-type material. In a p–n junction based on silicon, the combination of electrons and holes releases energy as heat that simply warms up the junction. However, if the p–n junction is made from a semiconductor such as gallium arsenide, energy escapes from the junction in a fairly narrow band at infrared wavelengths. And by mixing other elements in very small quantities with the gallium arsenide, visible light at other wavelengths is produced that escapes from the surface of the diode. The table below shows the wavelengths of the light emitted by different types of LEDs made from different semiconductors.

Semiconductor	Light	Wavelength (μm)
Gallium arsenide (GaAs)	infrared	0.9
Gallium arsenide phosphide (GaAsP)	red	0.65
Gallium phosphide (GaP)	green	0.56
Gallium indium phosphide (GaInP)	yellow	0.50
Indium gallium nitride (IGaN)	blue	0.40

As a rule of thumb, different colour LEDs require different forward voltages to operate them. Thus red LEDs take the least, and as the colour moves up the colour spectrum toward blue, the voltage requirement increases. Typically, a red LED requires about 2 volts, while blue LEDs require around 4 volts. Typical LEDs, however, require 20 to 30 mA of current, regardless of their voltage requirements. Light-emitting diodes last longer than their filament counterparts especially when operated at low currents and at low temperatures. High temperatures and currents will shorten their life. Quoted lifetimes are 25,000 to 100,000 hours.

A resistor must be connected in series with an LED if it is to be lit by a p.d. greater than, say, 2 V. Suppose the LED is to be lit by a 9 V battery as shown in Figure 7.15. The value of resistor R_1 is calculated as follows. Assume that the current through the LED is 20 mA and the p.d. across it is 2 V.

$$\text{p.d. across } R_1 = (9 \text{ V} - 2 \text{ V}) = 7 \text{ V}$$

$$R_1 = 7 \text{ V}/20 \text{ mA} = 350 \text{ }\Omega$$

Therefore, a suitable value for R_1 is 330 Ω. Higher values of R_1 will reduce the current through the LED and it will not light as brightly, but it will last longer before failing.

The small size, low energy consumption and low maintenance of LEDs makes them ideal as status indicators and as displays on a variety of equipment and installations. Large area LED displays are used for message and video displays in football stadiums, for electronic arrival and departure airport display signs and as dynamic decorative displays. Red or yellow LEDs are used in indicator and alphanumeric displays in environments where night vision must be retained: aircraft cockpits, submarine and ship bridges, astronomy observatories, and in the field, for example night-time animal watching and military field use.

Because of their long life and fast switching times, LEDs are used in car high-mounted brake lights. There are also design advantages to using LEDs in car headlights and rear lights, since LEDs are capable of forming much narrower beams of light than incandescent lamps with parabolic reflectors.

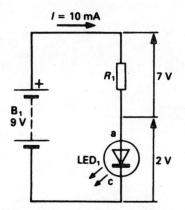

Figure 7.15 Calculating the value of the series resistor for an LED.

7.6 Power control with thyristors and triacs

Power control circuits make use of thyristors and triacs that, like rectifiers, depend on the properties of the p–n junction. They are to be found in food mixers, electric drills and lamp dimmers, for example. Figure 7.16a shows the appearance of a typical thyristor. This device was formerly known as a silicon controlled rectifier (s.c.r.) since it is a rectifier that controls the power delivered to a load, e.g. a lamp or motor. The symbol for a thyristor is shown in Figure 7.16b: it looks like a diode symbol with anode and cathode terminals, but with the addition of a third terminal called a gate, g. A simple demonstration of d.c. power control using the thyristor, Thy_1, is shown in Figure 7.16c. This circuit forward biases the thyristor but the thyristor does not conduct until a current is supplied to the gate terminal by closing switch SW_1. This small current flows into the gate terminal and the thyristor 'fires'. However, conduction continues and the lamp remains lit even when the current gate is removed by opening SW_1. The only way to switch the thyristor off is to open switch SW_2. Four layers of semiconductors make up the thyristor's p-n-p-n sandwich construction as shown in Figure 7.16d. Note that the word thyristor is derived from the Greek *thyra*, meaning door, and indicates that the thyristor is either open or closed.

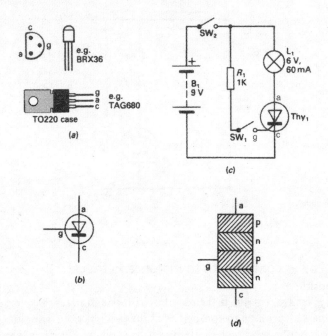

Figure 7.16 The Thyristor: (a) two types of package; (b) circuit symbol; (c) a simple power control circuit; (d) internal construction.

It is possible to use a thyristor to control a.c. power by allowing current to be supplied to the load only during part of each cycle. Figure 7.17 shows the basic circuit and waveforms. In a practical circuit, the gate pulses are applied automatically at a selected stage during part of each cycle. Thus, half power to the load is achieved by applying the gate pulses at the peaks of the a.c. waveform. More or less power in the load is achieved by changing the timing of the gate pulses.

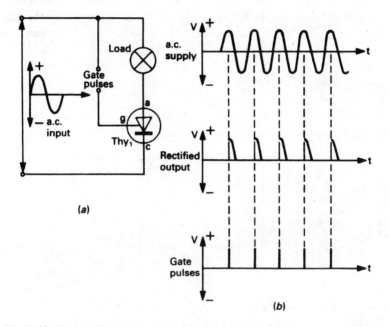

Figure 7.17 (a) Using a thyristor for a.c. power control, and (b) the waveforms that show how it works.

Since the thyristor switches off during the negative half cycles, it is only a half-wave device (like a rectifier) and allows control of only half the power available in a.c. circuits. A better device is a triac which comprises two thyristors connected in parallel but in opposition and controlled by the same gate, i.e. it is bi-directional and allows current to flow through it in either direction. Figure 7.18a shows its symbol. The terms anode and cathode have no meaning for a triac; instead the contact near the gate is called 'main terminal 1'(MT_1), and the other 'main terminal 2' (MT_2). As with the thyristor, a small gate current switches in a very much larger current through the main terminals. A typical gate current is of the order of 20 mA but it is adequate for triggering triacs of up to a rating of 25 A.

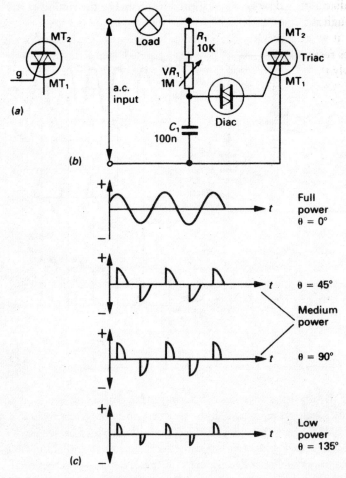

Figure 7.18 The triac: (a) circuit symbol; (b) a basic circuit for a.c. power control; (c) the waveforms that make it work.

If the triac is to be used in a lamp dimmer unit or a motor speed controller, there has to be some means of varying the a.c. power passing through the load. Figure 7.18b shows how this is achieved using an 'RC phase shifter'. VR_1 is the dimmer control. The device called a diac is in effect two zener diodes connected back to back. It conducts in either direction when the voltage across C_1 reaches the diac's breakdown voltage of about 30 V. The burst of current through the diac 'fires' the triac. As you now know (Chapter 6), the rate at which C_1 charges depends on the value of C_1 and that of the series resistor, VR_1 $(+R_1)$. The greater the value of VR_1, the more slowly the capacitor charges and the later in each half-cycle is the lamp turned on, and the dimmer the lamp.

The waveforms in Figure 7.18c show how the triac controls power by chopping off part of each half cycle. The amount chopped off is indicated by the phase shift, θ. If θ is zero, the triac conducts throughout the whole of each half-cycle of the a.c. waveform and the load is at full power. As θ increases from 0° to 180°, more and more of each half cycle is chopped off, and the dimmer the lamp becomes. Practical a.c. power control circuits using triacs refine the basic design in Figure 7.18b to give better low-level power control, and a reduction of the radio interference that is generated by the rapid turn-on and turn-off of the triac. You will not be surprised to discover that a quadrac combines the triac and diac in a single package.

TEST YOURSELF

1 Is the following statement true or false?

   Diodes are unlike ordinary resistors because they allow current to flow through them easily in one direction but not the other.

2 If a voltage is applied across the terminals of a rectifier diode so that current easily flows through it, the diode is said to be:
   (a) forward biased
   (b) reverse biased
   (c) unstable?

3 Which lamp lights in the circuits shown in Figure 7.19a and b below?

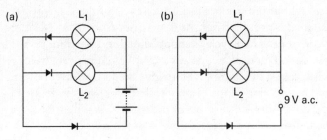

Figure 7.19

4 State two differences between the forward-biased and reverse-biased regions of a diode's V–I characteristic.

5 Is the following statement true or false?

   Half-wave rectification means that the whole of one half cycle of an alternating voltage is removed by a diode.

6 What is meant by 'ripple voltage' in a rectifier circuit? How is ripple voltage smoothed?

7 *Draw a diagram to explain how a diode is used to protect a d.c. circuit from connecting a battery to it the wrong way round.*

8 *Draw the V–I characteristic for a rectifier diode and for a zener diode.*

9 *Explain how a thyristor acts as an electronic switch.*

10 *How are triacs used for controlling a.c. power to a lamp?*

....

Amplifiers and transistors

In this chapter you will learn:
- *the meaning of voltage gain and power gain of an amplifier*
- *the use of the decibel in specifying voltage gain and power gain*
- *the meaning of bandwidth of an audio amplifier*
- *the operating principles of npn/pnp transistors and field effect transistors*
- *how transistors are used as switches and audio amplifiers*
- *how to sketch the connections of a Darlington pair transistor and describe the use as a switch*
- *how to use an audio amplifier in integrated circuit form.*

8.1 What amplifiers do

Figure 8.1 shows what an amplifier does. It increases the power of the signal passing through it. Thus a low-power signal, P_{in}, enters from the left, energy is drawn from the power supply, and a signal of higher power, P_{out}, leaves from the right. Note that an amplifier cannot increase the power of a signal without power being drawn from somewhere else, such as from a battery or some other source of electrical energy. In systems diagrams, it is usual to represent an amplifier black box as a triangle. In effect, this symbol is an arrowhead representing the direction in which the signal travels from the input to the output of the amplifier.

Now the ratio of the output power to the input power is the power gain, A, of an amplifier. Thus

$$\text{power gain} = \frac{\text{output power}}{\text{input power}}$$

$$\text{or } A = \frac{p_{out}}{p_{in}}$$

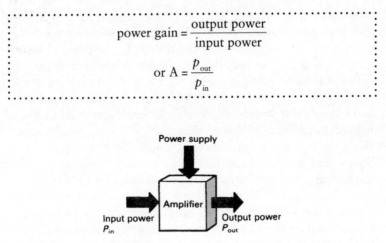

Figure 8.1 The function of an amplifier.

The power of electrical signals is measured in watts. Suppose the output power of an amplifier is equal to 50 W. If the input power is equal to 0.01 W (10 mW), the power gain of the amplifier is given by

$$A = \frac{50 \text{ W}}{10 \text{ mW}} = 5000$$

Thus, the output power from this amplifier is 5000 times more than the input power.

Amplifiers are very widely used in electronic systems, for example hi-fi amplifiers increase the power of audio frequency signals from compact discs before delivery to a loudspeaker. So let us take a closer look at the audio amplifier, which is perhaps the one you are most likely to come across.

8.2 Audio amplifiers

Audio amplifiers are to be found in hi-fi amplifiers, car radios, mobile phones, cassette players, MiniDisk players, hearing aids and satellite communications systems, to name but a few. In these systems, their purpose is to increase the power of audio frequency (AF) signals in the range of human hearing, which is between 20 Hz to 20,000 Hz (20 kHz). A 20 Hz audio frequency signal sounds like a 'buzz', and one of 10 kHz sounds like a high-pitched whistle. Few adults can hear AF signals above about 16 kHz, though many young children and animals do hear sounds that have frequencies higher than this. As we get older, our ears tend to become increasingly less sensitive to audio frequencies over 10 kHz.

The easy way to amplify a weak audio signal is to use an integrated circuit designed for the job. A good example of such an audio IC is the TBA820M device shown in Figure 8.2. A similar audio amplifier is the LM386 that features in Project 7 in Chapter 16. The TBA820M is housed in an 8-pin dual-in-line package that protects the 2 mm by 2 mm square silicon chip on which the amplifying circuit is made. When it is operated from a 12 V power supply, this device provides a maximum audio power of 2 W from

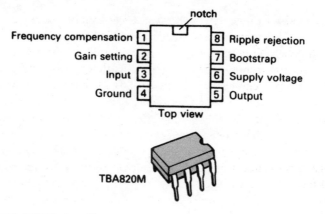

Figure 8.2 The TBA820M audio amplifier.

an 8 ohm loudspeaker. It is actually designed to operate from any power supply voltage in the range 3 V to 16 V, though at 3 V its power output is much less than at 16 V. The manufacturers of this audio amplifier have also made sure that when it is not actually amplifying a signal it draws very little current from the power supply – it is said to have a low quiescent current.

8.3 The bandwidth of an amplifier

Most audio amplifiers are designed to amplify all frequencies within a specified band of frequencies by the same amount. This is explained by means of the frequency response graph shown in Figure 8.3. The level part of the graph is typical of the 'flat' frequency response of a hi-fi amplifier. However, the power gain of most audio amplifiers falls off sharply at frequencies below about 20 Hz and above about 20 kHz. Surprisingly, the power gain of some good-quality audio amplifiers only begins to fall off at frequencies above 40 kHz. Manufacturers claim that this ability to amplify frequencies well above the range of human hearing improves the sound reproduction within the audio frequency range.

The frequency range over which the power gain does not fall below half the maximum power gain is called the bandwidth of the amplifier. Note that frequencies are plotted 'logarithmically' on the frequency response graph. This means that ten-fold increases of frequency occupy equal distances along the horizontal axis. Thus, there is equal spacing between 10 Hz and 100 Hz, and between 100 Hz and 1000 Hz, and so on, so that the graph can accommodate the wide range of frequencies and show clearly the amplifier's frequency response in the 20 Hz to 20 kHz frequency range. If the calibrations along the frequency axis were linear, this 100-fold frequency range would occupy only about one-tenth of the full bandwidth of 20 kHz. As it is, it occupies about one-half of the 20 kHz range.

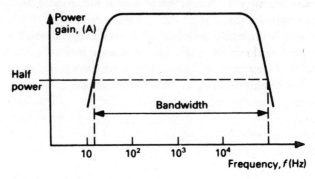

Figure 8.3 The frequency response graph of an audio amplifier.

Though a good quality amplifier has a flat frequency response over a wide range, it is common for manufacturers to provide a means to adjust this response to suit individual tastes. This may take the form of a single potentiometer, a 'tone control' that can be adjusted to select relatively more high or low audio frequencies. Or the device might be equipped with a 'graphic equalizer', which allows control of the frequency response in the low, middle and upper frequency range of the audio band.

8.4 The decibel

The decibel (dB) is often used as a measure of the loudness of sounds, e.g. the sound from a low-flying jet at an airport might be 60 dB. It is also used when talking about the power gain of an amplifier. The decibel actually derives from the bel that was introduced in the early days of telephone engineering to express how much more power one signal had than another. In practice, the bel turned out to be too large a unit so the decibel (one-tenth of a bel) is now used.

Insight

The bel is named in honour of Alexander Graham Bell, the pioneer of telephones. Born in Scotland in 1847, he and his family were authorities in elocution. As a teenager, he began to research the mechanics of speech with the aim of making a device that could mimic the human voice. The onset of tuberculosis caused the family to move to Canada where he could recuperate and where eventually he stayed. The idea of transmitting speech along a wire never left him. The final breakthrough came about while testing a circuit with one transmitter and two receivers on 2 June 1875. The transmitter was switched off and Watson, his assistant, was adjusting one of the receivers when Bell heard a note coming from the receiver in his room. With the transmitter turned off, the note had to be coming from the other receiver. He had discovered that the receiver could also work in reverse: instead of making sound when electricity was sent through it, it also made electricity when supplied with sound because the sound moved the magnet in the coil and generated electricity. Developments were swift. Within a year the first telephone exchange was built in Connecticut and within the decade more than 150,000 people in the USA alone owned telephones. The Bell Telephone Company was created in 1877, with Bell the owner of a third of the 5000 shares.

The reason that the decibel is such a useful unit for comparing the strengths of audio frequency signals is that the response of the ear is logarithmic. This means that to the human ear a change in loudness is the same whether the sound power increases from 0.01 W to 0.1 W, or from 0.1 W to 1 W. Thus, ten-fold increases in sound power seem like equal increases in loudness. Actually, the ear does not respond like this over the full range of audio frequencies, for it is more responsive to some audio frequencies than to others, being most sensitive at about 4 kHz.

Suppose that an amplifier increases the power of a signal from P_{in} to P_{out}. Then the power gain is the ratio P_{out}/P_{in}. The following equation expresses this gain in decibels.

$$\text{power gain} = 10 \times \log_{10}\left(P_{out}/P_{in}\right)$$

For example, if the input power is 0.1 W and the output power is 10 W, the power gain is

$$10 \times \log_{10} 10 \text{ W}/0.1 \text{ W}$$
$$= 10 \times \log_{10}(100)$$
$$= 10 \times 2 = 20 \text{ dB}$$

As a result, a 10-fold increase in power can be expressed as a 20 dB increase in power. A similar calculation shows that a 1000-fold increase in power is equivalent to a 30 dB increase; a million-fold increase to a 60 dB increase; and so on. Each ten-fold increase of power gain is equivalent to a 10 dB increase. Figure 8.4 shows that the decibel scale is a compact way of representing a wide range of power gains.

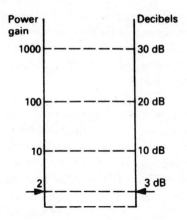

Figure 8.4 A decibel scale for power gain.

Manufacturers of audio amplifiers often use the decibel scale to specify the bandwidth of their amplifiers. For example, Figure 8.5 is the frequency response graph of Figure 8.3 redrawn to show the gain on the vertical axis in decibels. Now the bandwidth is shown as 20 Hz to 20 kHz between the 3 dB points. But why 3 dB?

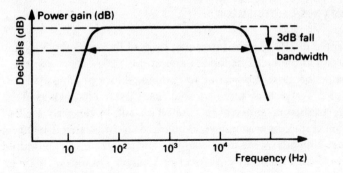

Figure 8.5 The 3dB points.

Well, as you now know, the bandwidth of an amplifier is defined between the points on the frequency response graph where the power has fallen to half its 'flat' value. So the fall in gain expressed in decibels is $10 \times \log_{10}(1/2) = -3$ dB. The minus sign is important for it indicates a fall in power. In terms of the decibel, a doubling in power is equivalent to an *increase* of 3 dB, and a halving in power to a *decrease* of −3 dB.

Actually, it is easier to determine the gain of an amplifier by measuring the values of the p.d.s generated across input and output resistances by the input and output signals, rather than by measuring the input and output powers. In this case, the power gain of the amplifier can be found from the following equation:

$$\text{power gain} = 20 \log_{10}(V_{out}/V_{in})$$

where V_{in} is the p.d. generated across the input resistance by the input signal, and V_{out} is the p.d. generated across the output resistance by the output signal. The equation assumes that the input and output resistances are the same.

8.5 Types of transistor

Invented in 1947 by the three-man team of Bardeen, Shockley and Brattain at the Bell Telephone Laboratories, USA, the transistor became the most important basic building block of almost all circuits. Transistors were first used singly in circuits such as in the early 'transistor' radios of the late 1950s but, by the early 1970s, silicon chips comprising several hundred transistors were being made. Nowadays, the most complex integrated circuits contain upwards of a million transistors (see Chapter 12).

Transistors are still used as discrete, i.e. individual, components in circuits. Figure 8.6 shows the distinguishing features of a small selection of transistors that fall into two main categories depending on the way n-type and p-type semiconductors are used to make them. One sort is the bipolar transistor (the sort invented in 1947) of which there are two types, npn and pnp, as shown by the examples in Figure 8.6a to c. The second sort is the unipolar transistor (a later invention) that is also called the field-effect transistor (FET). Examples of the FETs, of which there are two sorts, n-channel and p-channel, are shown by the examples in Figure 8.6d to f. Bipolar transistors have three leads called emitter (e), base (b) and collector (c); and the leads of an FET are source (s), gate (g) and drain (d). These transistors are housed in a metal or plastic package. In many of the 'metal can' types, e.g. the TO18 shape, the collector lead is also connected to the metal can. Transistors used for high power applications have a flat metal side to them, e.g. the TO220 shape, to which a heat sink can be bolted to help the transistor dissipate excess heat produced within it. The symbols for bipolar and unipolar transistors are shown in Figure 8.7. The npn bipolar transistor and the n-channel field-effect transistor are the more commonly used and it is these two that we shall look at more closely, first as electronic switches and then as audio amplifiers.

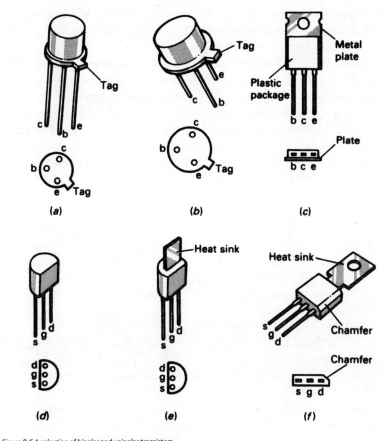

Figure 8.6 A selection of bipolar and unipolar transistors.
In TO18 outline, (a) npn types BC108 and 2N2222; and pnp types BC478 and 2N2907;
in TO39 outline, (b) npn type 2N3019 and pnp type 2N2905;
in TO 220 outline, (c) npn, type TIP31A and pnp type TIP32A;
in TO92(D) outline, (d) n-channel junction gate FET, type BF245C;
in TO237 outline, (e) n-channel metal-oxide field-effect transistor (MOSFET), type VN10LM;
in TO202(B) outline, (f) n-channel MOSFET, type VN46AF.

Insight

When John Bardeen, Walter Brattain and William Shockley invented the transistor at Bell Labs in 1947, no one really knew quite what to do with the invention. But once circuit designers realised that the tiny switches would enable products to be built smaller, more reliably and with less

(Contd)

power consumption than with conventional electronic valves, the market started to develop. So five years later the first practical application appeared – a hearing aid. Now this was not a large market but it did demonstrate two key features of the new technology: it enabled something to be done that was impossible with the previous valve technology and it helped enhance the quality of life for those in need. These two features, along with the industry's ability to reduce the production cost of a transistor by around 30% per year, have subsequently proved to be the fundamental drivers behind today's enormously profitable electronics industry.

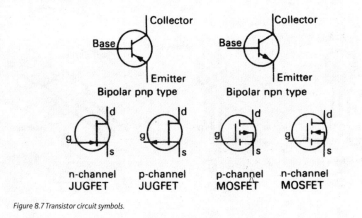

Figure 8.7 Transistor circuit symbols.

8.6 How bipolar transistors work

Figure 8.8a shows a simple model of an npn bipolar transistor. The base terminal connects to a very thin layer of p-type semiconductor that is sandwiched between two layers of n-type semiconductor. This npn transistor comprises two p–n junctions, and it is useful to think of it as made up of two diodes connected together as shown in Figure 8.8b. When an npn transistor is acting as a switch or an amplifier, the direction of the (conventional) current flow into and out of its terminals is shown in Figure 8.8c. A small current into the base terminal causes a much larger current flow between the emitter and collector terminals. Thus, an npn transistor amplifies current. Note that when current flows through the npn transistor, the base–emitter p–n junction is forward-biased and the base-collector junction is reverse-biased.

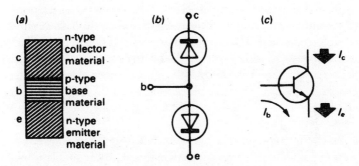

Figure 8.8 An npn bipolar transistor: its (a) structure, (b) diode equivalent circuit, and (c) operating currents.

The ratio of the collector current, I_c, to the base current, I_b, is known as the current gain of the transistor and is represented by the symbol, h_{FE}, so that

$$h_{FE} = \frac{\text{collector current}}{\text{base current}} = \frac{I_c}{I_b}$$

The current gain varies greatly from transistor to transistor, and even between transistors of the same type. It may be as little as 10 for a high-power transistor such as one used in the output stages of a hi-fi amplifier, to as much as 1000 for a low-power transistor used in the input stages of a hearing aid. Fortunately, in most circuit designs it is possible to compensate for individual variations in current gain. A manufacturer merely guarantees that a particular device has a gain in a specified gain band. Note the relationship between the collector, emitter and base currents:

$$I_e = I_c + I_b$$

since the current flowing into the transistor must equal the current flowing out of it. If the current gain is large, e.g. more than 100, I_b is much less than I_e or I_c, so $I_c = I_e$, approximately.

The way an npn transistor amplifies current is quite complicated and rather difficult to understand, but a simple model should help you to grasp the general principles. Figure 8.9a shows the distribution of electrons and holes in an npn transistor when the base terminal is left unconnected. The external power supply makes the collector terminal positive with respect to the emitter terminal. This external power supply reverse biases the p–n junction formed by the collector and base regions of the transistor. Thus, electrons and holes move away from the collector–base junction and no current flows across it.

Figure 8.9b shows what happens when the npn transistor is working as a current amplifier. An external power supply now forward biases the base–emitter junction, and electrons flow easily across this junction. A few of these electrons combine with holes in the p-type base region. Not many electrons 'get lost' in this way, for two reasons: first, the base region is lightly doped with p-type impurities (Chapter 2) and, second, the base region is very thin. Making up for the 'loss' of electrons by recombination in the base region, is an equal number of electrons leaving the base terminal, and these comprise the base current, I_b. Since the base region is very thin, most of the electrons move rapidly across the emitter–base junction and are swept over the reverse-biased collector–base junction forming the collector current, I_c. In effect, the transistor has amplified the small base current to produce a larger collector current. Thus if 0.2 per cent of the electrons that moved across the emitter–base junction combine with holes in the base, and the emitter current is 100 mA, the base current is

$$(0.2/100) \times 100 \text{ mA} = 0.2 \text{ mA}$$

Thus, the current gain of the transistor is

$$I_c/I_b = (I_e - I_b)/I_b$$
$$= (100 - 0.2) \text{ mA} / 0.2 \text{ mA}$$
$$= 500$$

Note that an npn transistor amplifies only when the base–emitter junction is forward biased, i.e. when the base terminal is more positive than the emitter terminal by about 0.6 V. The reason why the npn transistor is called a 'bipolar transistor' is that both electrons and holes are responsible for its operation. Also, note that practical transistor circuits show only conventional current flow into and out of the transistor. Thus, the arrow inside the npn transistor symbol shows the conventional current flow across the forward-biased base–emitter junction of the transistor.

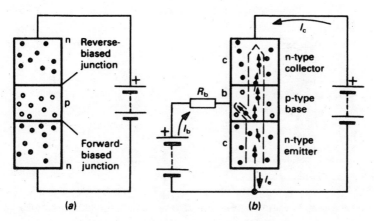

Figure 8.9 Electrons and holes in a bipolar transistor: (a) transistor is switched off, and (b) transistor is switched on.

8.7 How field-effect transistors work

The field-effect transistor (FET) has some characteristics that make it a better choice than the bipolar transistor in electronic switches and amplifiers. These are its lower power consumption, its higher input resistance, and its higher frequency response. The lower power consumption makes it a good choice for integrated circuits, since excessive heat is not produced when many thousands of FETs are integrated on a small area of a silicon chip (Chapter 12). It is therefore possible to design portable battery-operated devices, e.g. personal computers and solar-powered calculators, using this type of integrated circuit. Furthermore, FETs take up less space than bipolar transistors so more of them can be packed together on a silicon chip. In discrete form, FETs look much the same as bipolar transistors but they work quite differently. There are two main types of FET, the junction-gate FET (JUGFET) and the metal-oxide semiconductor FET (MOSFET). There are n-channel and p-channel versions of each type. It is the n-channel version of these FETs that is described here, since it is more widely used than its p-channel counterpart. Let us look at the basic structure of each type of FET, the JUGFET and the MOSFET.

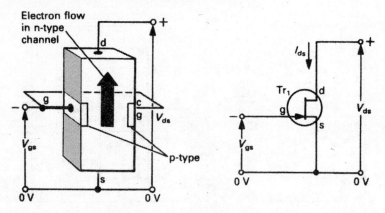

Figure 8.10 Current flow through an n-channel junction-gate field-effect transistor (JUGFET).

Figure 8.10 shows the basic structure and circuit symbol of the JUGFET. It comprises a channel of n-type semiconductor through which electrons flow between the source terminal, s, and the drain terminal, d. The gate-to-source voltage, V_{gs}, controls this current, I_{ds}. Note that the gate terminal is connected to a p-type semiconductor that is embedded in the n-type channel. Since the gate terminal is more negative than the source terminal, the p–n junction between the p-type gate and the n-type channel is reverse-biased and this causes a depletion region to extend into the n-type channel. Since the depletion region contains a negligible number of electrons (Chapter 7), electrons can only flow in what is left of the n-type channel. The resistance of this channel, and hence the size of I_{ds}, is determined by the width of the channel. And this width depends on the gate-to-source voltage, V_{gs}.

Figure 8.11 compares the width of the n-type channel for two different values of V_{gs}. The more negative V_{gs} is, the narrower the channel, the larger the channel resistance and the smaller I_{ds}. When V_{gs} has reached what is called the 'pinch-off voltage', the depletion region completely closes the channel and the resistance of the channel is very high. The pear-shaped structure of the depletion region shows that the depletion region is thicker at the drain end of the channel since here the p.d. between the drain and gate is greater.

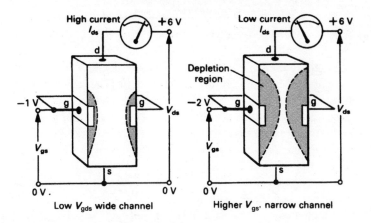

Figure 8.11 How the gate-to-source voltage controls the channel width in a JUGFET.

Note the main difference between the operation of the npn bipolar transistor and the n-channel JUGFET: the bipolar transistor depends on the flow of both electrons and holes (hence 'bi-polar'), but the n-channel JUGFET relies on the flow of electrons only, i.e. it is a unipolar device. (The p-channel FET relies on the flow of holes only.) Furthermore, whereas it is essential that a small flow of electrons leaves from the base terminal of an npn transistor, a negligible and non-essential current leaves the gate terminal of an n-channel JUGFET. Thus, the FET controls the flow of electrons through the n-channel by the electric field established between the gate terminal and the channel (by V_{gs}). That is why the 'field-effect transistor' is so-called. Consequently, it has an input resistance that is considerably higher than that of a bipolar transistor, and this is important in applications where a signal source such as a crystal microphone provides changes of voltage without the capacity to deliver much current.

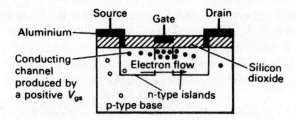

Figure 8.12 The structure of an n-channel enhancement MOSFET.

Now compare the structure of the n-channel JUGFET with that of the n-channel MOSFET shown in Figure 8.12. The structure has a gate terminal that is insulated from the n-type channel by a layer of electrically insulating silicon dioxide. This means that almost no current, even less than for a JUGFET, can flow to or from the gate terminal. There are actually two types of n-channel MOSFET, the depletion MOSFET and the enhancement MOSFET. In the latter device, a positive value of V_{gs} actually produces, or enhances, an electron-rich n-channel just below the oxide layer. Consequently, the n-channel enhancement MOSFET is normally off until it receives a positive V_{gs}. Changes to V_{gs} cause changes to the electron density in this channel and hence changes in I_{ds}. While its operation is basically the same as an n-channel JUGFET, the n-channel MOSFET requires a positive V_{gs} and the JUGFET a negative V_{gs}. The symbols for the MOSFETs shown in Figure 8.9 are for the enhancement types. These n-channel and p-channel MOSFETS are also called n-MOSFETs and p-MOSFETs.

8.8 Transistors as electronic switches

Despite the advantages of using integrated circuits in circuit design, millions of individual transistors are made each year, some of which are used in simple control circuits like the one shown in Figure 8.13. This circuit switches on the lamp L_1 when the light-dependent resistor LDR_1 is covered as it is a dark-operated switch. Similar circuits are used in automatic streetlights that come on at dusk. This electronic switch comprises three building blocks: a sensor based on resistors R_1 and LDR_1 that make up a potential divider; a current amplifier based on transistor Tr_1, and an output device, in this case the lamp L_1. The voltage at point X controls the on/off action of this circuit. This voltage rises when LDR_1 is covered, and falls when LDR_1 is illuminated. When the voltage at X rises, the current through the lamp increases; when it falls the lamp current decreases.

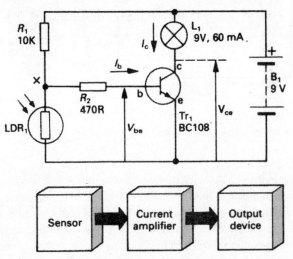

Figure 8.13 Single transistor switching circuit comprising three building blocks.

Figure 8.14a shows the action of the npn transistor as a current amplifier; small changes of base current produce large changes in the collector current. The ratio I_c/I_b is the current gain, 250 in this case. But I_b and hence I_c is controlled by the base–emitter voltage, V_{be}, as shown in Figure 8.14b, a graph known as a transfer characteristic. If V_{be} is less than 0.6 V, I_b and I_c are both zero, and the lamp is off. In this state, the collector–emitter resistance is very high and the collector–emitter voltage V_{ce}, is +9 V. The transistor is said to be 'cut off' and the transistor switch is 'open'. As V_{be} increases above 0.6 V, I_b increases, and the amplifying action of the transistor produces a much larger collector current, I_c. The lamp brightens until V_{be} has reached about 0.75 V. In this state, the collector–emitter resistance is so low that V_{ce} is nearly zero.

The transistor is then said to be saturated since any further increase in V_{be} does not increase I_c. The transistor switch is 'closed'.

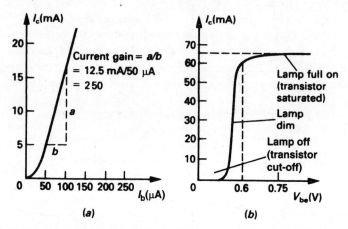

Figure 8.14 Characteristics of an npn transistor: (a) variation of collector current with base current; (b) variation of collector current with base voltage.

One of the drawbacks with this simple one-transistor switch is that there is a small range of illumination of LDR_1 over which the lamp is only partially switched on. However, if two transistors are used as shown in Figure 8.15, the 'sharpness' of the switching action is improved considerably. This circuit is known as a Darlington pair. In this design, the emitter current of Tr_1 provides the base current of Tr_2. Also note that the voltage on the base of the first transistor must now exceed 2×0.6 V = 1.2 V before lighting the lamp. The overall current gain of this combination of two transistors is the product of their individual current gains. That is, if the gain of each transistor is 100, the overall current gain is $100 \times 100 = 10,000$. At least, this is the theoretical value, but for practical reasons the overall gain may be somewhat less than this. The collector current of Tr_1 is a hundred times smaller than the collector current of Tr_2 so that most of the current flowing through L_1 flows through Tr_2. The overall effect is that the base current of Tr_1 is a hundred times smaller than when one transistor is used. This makes the circuit much more sensitive than the single transistor circuit, and lamps or relays are switched on and off smartly for small changes of light intensity.

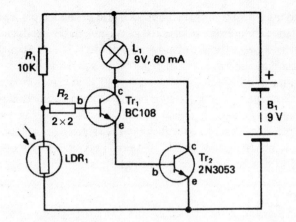

Figure 8.15 A Darlington pair transistor switch.

Darlington pair transistors are available in discrete packages having just three leads: the base of the first transistor, the common collector lead and the emitter of the second transistor. Indeed, they are also available in integrated circuit packages as shown in Figure 8.16. This arrangement is ideal when computers are used to control several motors or lamps via relays. The Darlington pairs amplify the low current signals available from a computer's output lines.

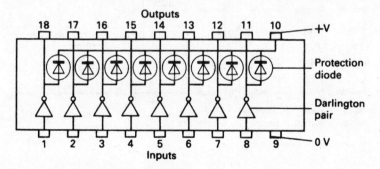

Figure 8.16 Darlington pairs in an integrated circuit package.

Figure 8.17 shows the sensors that can be used in a Darlington-pair switching circuit. The variable resistor VR_1 enables some of the circuits to be set to switch at a predetermined level. Thus a 'rain alarm' is obtained using sensor S_1 that comprises a grid of conductors separated by insulation. If moisture bridges the gaps between the conductors, the relay energizes. The use of a thermistor, Th_1 (Chapter 5), turns the circuit into a thermostat that could be used to maintain the temperature of a tropical fish tank at, say, 30°C. A proximity switch is obtained using a reed switch that is operated magnetically (Chapter 4). Note that if any one of these sensors and the resistor in series with it change places, the circuits have the opposite switching function. Thus, if LDR_1 is interchanged with VR_1, the circuit becomes a light-operated switch and switches on the relay when LDR_1 is illuminated.

Insight

The arrangement of two transistors as a Darlington pair was invented by Sidney Darlington in 1953. He patented the idea of connecting two transistors on a single chip sharing a common collector in order to provide an enhanced current gain.

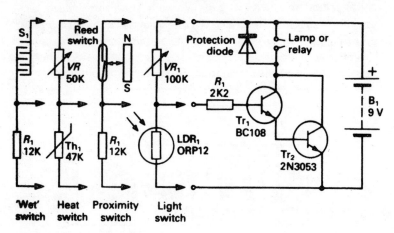

Figure 8.17 Options for using a Darlington pair.

Incidentally, note that in transistor switching circuits that energize a relay, a silicon diode is connected across the relay. The diode is reverse biased relative to the power supply; its anode terminal is connected to the collector terminal. Its purpose is to protect the transistor from possible damage due to the high back e.m.f.

generated by the relay when it de-energizes. Since the diode is reverse biased, it causes no drain on the power supply. However, the diode harmlessly short-circuits the back e.m.f. produced when the relay de-energizes since this is directed in the forward bias direction, i.e. from collector to power supply, anode to cathode.

Figure 8.18 shows the general shape of the transfer characteristic of a MOSFET device such as the VN10LM or VN46AF (see Figure 8.6). This graph is quite different from the transfer characteristic (Figure 8.14b) of a bipolar transistor. The npn bipolar begins to conduct when its base–emitter voltage is about 0.6 V, and an increase of about 0.2 V is sufficient to saturate the transistor. However, the gate-to-source voltage, V_{gs}, required to make a MOSFET device begin to conduct is much more variable, usually being between about 0.8 V and 2 V (known as the 'threshold voltage'). Furthermore, the value of V_{gs} needed to bring the device into saturation is very much higher than this, generally being in the region of about 10 V. Thus, the drain current continues to increase over a wide range of gate voltage. In this respect, the MOSFET device is inferior to the bipolar transistor as a switch. However, it does have one great advantage: whereas a high-power bipolar transistor might require a base current of, say, 100 mA to bring it into saturation, a MOSFET device requires virtually no current at all! It is voltage-operated, not current-operated.

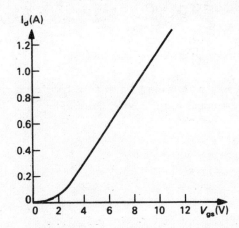

Figure 8.18 The transfer characteristic of a MOSFET.

TEST YOURSELF

1   Write down the equations for calculating the power gain of an audio amplifier.

2   An amplifier increases the power of a signal from 10 mW to 100 W. What is its power gain:
    (a)   as a number?
    (b)   as decibels?

3   An amplifier produces a power gain of 20 dB. What is the ratio of the input to the output power?

4   How is the bandwidth of an amplifier defined? Is it:
    (a)   The highest frequency it amplifies?
    (b)   The length of the tuning scale on a radio receiver?
    (c)   The frequency range where the voltage falls by no more than 0.7 of the maximum gain?

5   Distinguish between npn and pnp transistors in terms of their structure and circuit symbol.

6   The base region of an npn transistor is made of ..... semiconducting material.
    p-type / n-type / thick?

7   Is the following statement true or false?

    When a transistor is biased normally, the base current is always greater than the emitter current.

8   State two advantages of the Darlington pair circuit compared with using just one transistor.

9 *Under certain conditions, a bipolar transistor requires a base current of 4 mA to produce a collector current of 340 mA. What is the d.c. current gain of the transistor under these conditions.*

10 *A field-effect transistor has a [high/low] input resistance compared with a bipolar transistor. (Delete as appropriate.)*

....

Logic gates and Boolean algebra

In this chapter you will learn:
- *to explain the meaning and purpose of a 'truth table' for predicting the response of a logic gate to digital signals*
- *the truth tables for AND, NAND, OR, NOR, NOT and exclusive-OR logic gates*
- *to use Boolean algebra to represent the functions of the above logic gates*
- *why the NAND gate is called a universal logic gate*
- *the characteristics of two major digital logic families*
- *how logic gates can be used to solve simple logic problems.*

9.1 What logic gates do

The basic building blocks that make up all digital systems are simple little circuits called logic gates. A logic gate is a decision-making building block that has one output and two or more inputs as shown in Figure 9.1. Two examples of these decision-making logic gates were introduced in Chapter 3. These were the AND gate and the OR gate using mechanical switches. In this chapter, logic gates are contained within integrated circuits that operate electronically instead of mechanically. The input and output signals of these gates can have either of two binary values, 1 or 0. The binary value of the output of a gate is decided by the values of its inputs, and a truth table for a logic gate shows the value of the output for all possible values of the inputs. Like the other gates below, AND and OR gates are known as combinational logic gates. This term arises because their outputs are the logical (i.e. predictable) result of a particular combination of input states.

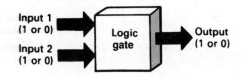

Figure 9.1 A logic gate has two or more inputs and one output.

Logic gates are used in many types of computer, control and communications systems, and especially in calculators and digital watches, and other devices that have digital displays. Figure 9.2 shows the usual place of logic gates in digital systems. They have the intermediate task of receiving signals from sensors such as keyboard switches and temperature sensors, making decisions based on the information received, and sending an output signal to a circuit that provides some action, such as switching on a motor, activating a 'go' or 'stop' lamp. Although logic gates can be designed using individual diodes and transistors, or even mechanical switches as explained in Chapter 3, most digital circuits now make use of logic gates in integrated circuit packages. There are two main 'families' of these digital ICs; one is called transistor–transistor logic (TTL), and the other complementary metal-oxide semiconductor logic (CMOS). Each family has its variants as explained in Section 9.5 below.

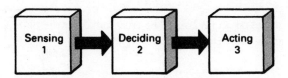

Figure 9.2 The function of a digital system: logic gates are used in building block 2.

9.2 Symbols and truth tables of logic gates

Two alternative systems are in use for showing the symbols of logic gates in circuit diagrams, the American 'Mil Spec' system and the British Standards system. Figure 9.3 summarizes these symbols for six logic gates, the AND, OR, NOT, NAND, NOR and exclusive-OR gates. The American Mil Spec symbols are widely preferred, since their different shapes are easily recognized in complex circuit diagrams although in the UK there is some pressure to adopt the British Standards symbols.

The AND gate gives an output of logic 1 when all inputs are at logic 1, and an output of logic 0 if any or all inputs are at logic 0. Therefore, an AND gate is sometimes called an 'all-or-nothing gate'. For the two-input AND gate shown in Figure 9.3a, the output, S, is at logic 1 only when input A and input B are at logic 1. The truth table for the two-input AND gate gives the state of the output, S, for all combinations of input states – hence the term 'combinational logic' is used to describe logic systems using gates like the AND gate.

For the two-input OR gate shown in Figure 9.3b, the output, S, is at logic 1 when either input A or input B, or both inputs, are at logic 1. Thus, the OR gate is sometimes called an 'any-or-all' gate. The truth table for the two-input OR gate gives the state of the output, S, for all combinations of input states.

For the NOT gate shown in Figure 9.3c, the output, S, is simply the inverse of the input A. Thus, if the state of the input is logic 1, the output state is logic 0 and vice versa. For this reason the NOT gate is also called an inverter. The truth table for the NOT gate is simple – it has only two lines.

The NAND (or NOT-AND) gate gives an output that is the inverse of the AND gate. Thus, for the two-input NAND gate shown in Figure 9.3d, the output, S, is at logic 0 when both input A and input B are at logic 1. The truth table for the two-input NAND gate gives the state of the output, S, for all combinations of input states. Compare it with the truth table of the AND gate.

The NOR (or NOT-OR) gate gives an output that is the inverse of the OR gate. Thus for the two-input NOR gate shown in Figure 9.3e, the output, S, is at logic 0 when either input A or input B, or both inputs, are at logic 1. The truth table for the two-input NOR gate gives the state of the output, S, for all combinations of input states.

The two-input Exclusive-OR (or XOR) gate shown in Figure 9.3f does something that the OR gate of Figure 9.3b does not: it is a true OR gate for it only gives an output of logic 1 when either, but not both, of its inputs is at logic 1. The truth table summarizes the action of the XOR gate and you should compare it with that for the OR gate.

Insight

In the binary number system, there are just two digits, 0 and 1, and binary numbers have weightings of 2^0, 2^1, 2^2, 2^3, and so on. A binary digit, 0 or 1, is known as bit, a name that is probably a contraction of binary digit – or possibly it means a small piece of information! In computing jargon, a binary number of four bits, e.g. 1101, makes a nibble, and eight bits, e.g. 11010010, makes a byte. Thus, the binary number $(1101)_2$ having a base of 2 is made up as follows:

digit	weighting	digit × weighting	decimal number
1	2^3	1×2^3	8
1	2^2	1×2^2	4
0	2^1	0×2^1	0
1	2^0	1×2^0	1
			13_{10} (sum)

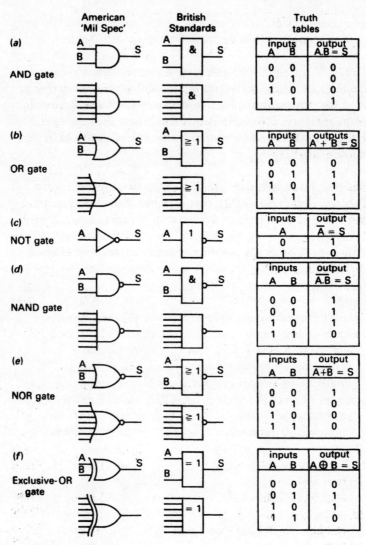

Figure 9.3 Symbols and truth tables for logic gates.

9.3 Introducing Boolean algebra

In 1847, George Boole invented a shorthand method of writing down combinations of logical statements that are either 'true' or 'false'. Boole proved that the binary or two-valued nature of his 'logic' is valid for symbols and letters as well as for words. His mathematical analysis of logic statements became known as Boolean algebra.

Since digital circuits deal with signals that have two values, Boolean algebra is an ideal method for analysing and predicting the behaviour of these circuits. For example, 'true' can be regarded as logic 1, i.e. an 'on' signal, and 'false' as logic 0, i.e. an 'off' signal. However, Boolean algebra did not begin to have an impact on digital circuits until 1938 when Shannon applied Boole's ideas to the design of telephone switching circuits.

The table below shows how Boolean algebra represents the functions of the five logic gates, AND, OR, NOT, NAND, NOR and XOR, discussed above.

Logic statement	Boolean expression
(a) AND gate Input A and input B = output S	$A \cdot B = S$
(b) OR gate Input A or input B = output S	$A + B = S$
(c) NOT gate Not input A = output S	$\overline{A} = S$
(d) NAND gate Not input A and input B = output S	$\overline{A \cdot B} = S$
(e) NOR gate Not input A or input B = output S	$\overline{A + B} = S$
(f) Exclusive-OR gate Input A or input B = output S (excluding input A and input B)	$A \oplus B = S$

Note the use of the following conventions in Boolean algebra:

(a) *two symbols are ANDed if there is a 'dot' between them;*
(b) *two symbols are 'ORed' if there is a '+' between them;*
(c) *a bar across the top of a symbol means the value of the symbol is inverted.*

The use of the bar across the top of a symbol is very important in Boolean algebra. Since A and B can take values of either 0 or 1, $\bar{1} = 0$ and $\bar{0} = 1$ as in the NOT gate. If the bar is used twice, this represents a double inversion. Thus, $\bar{\bar{1}} = 1$, and $\bar{\bar{0}} = 0$. What do you think happens to the value of the symbol if the bar is used three times?

Insight

George Boole (1815–64) was born in the English town of Lincoln and was lucky enough to have a father who passed on his own love of mathematics. Boole was translating Latin poetry by the age of twelve and soon became fluent in German, Italian and French. At 16 he became an assistant teacher, and at 20 he opened his own school.

Boole soon began to see the possibilities for applying his interest in maths to the solution of logical problems. Boole's 1847 paper, 'The Mathematical Analysis of Logic', argued that logic was principally a discipline of mathematics, rather than philosophy. Boole began to delve deeper into his own work, concentrating on refining his 'mathematical analysis', and determined to find a way to encode logical arguments into a language that could be manipulated and solved mathematically. He came up with a type of linguistic algebra, the three most basic operations of which were (and still are) AND, OR and NOT. He argued that these three functions are the only operations necessary to perform comparisons or basic mathematical functions.

9.4 The universal NAND gate

It is possible to use one or more NAND gates to produce the logic functions of AND, OR, NOT, NOR and XOR. That is why a NAND gate is called a 'universal logic gate'. The NOR gate has the same property but is used less that the NAND gate. Boolean algebra can be used to show that a NAND gate is 'universal' in this way.

To make a NOT gate, suppose we connect together the two inputs of a NAND gate as shown in Figure 9.4a, and apply a logic 1 to these two inputs. Now for a NAND gate, we can write the Boolean expression $\overline{A \cdot B} = S$. But $A = B = 1$, so $\overline{1.1} = \overline{1} = 0$. And if $A = B = 0$, $\overline{0.0} = \overline{0} = 1$. So a NOT can be obtained by connecting together the two inputs of a 2-input NAND gate and it then inverts the input signal. In fact, it does not matter how many inputs the NAND gate has: if all its inputs are connected together, the result is a NOT gate.

To make an AND gate, we need two NAND gates connected as shown in Figure 9.4b where the NAND gate is followed by a NOT gate. The output of the NAND (not-AND) gate is followed by an inverter, a NOT gate, to produce an AND gate. Thus, if R is the output from the first NAND gate, $R = \overline{A \cdot B}$. Hence $S = \overline{R} = \overline{\overline{A.B}}$, or $A.B = S$. The two bars signify a double inversion leaving the expression unchanged. The following truth table summarizes the action of this combination of NAND gates.

A	B	$\overline{A.B} = R$	$S = \overline{R}$
0	0	1	0
0	1	1	0
1	0	1	0
1	1	0	1

Note that the last column is the same as that of an AND gate (Figure 9.3a).

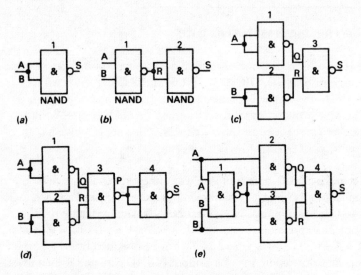

Figure 9.4 How NAND gates are used to produce other logic functions: (a) NOT gate: $S = \overline{A}$ (b) AND gate: $S = A.B$ (c) OR gate: $S = A + B$ (d) NOR gate: $S = \overline{A + B}$ (e) XOR gate: $S = A \oplus B$.

To make an OR gate, three NAND gates are needed as shown in Figure 9.4c. NAND gates 1 and 2 are each connected as NOT gates. Thus, inputs Q and R to NAND gate 3 are: $Q = \overline{A}$ and $R = \overline{B}$. Thus $\overline{\overline{A}.\overline{B}} = S$. The truth table for this combination is as follows:

A	$\overline{A}$	B	$\overline{A}$	$\overline{A}.\overline{B}$	$\overline{\overline{A}.\overline{B}}$
0	1	0	1	1	0
0	1	1	0	0	1
1	0	0	1	0	1
1	0	1	0	0	1

Note that the function of an OR gate is represented by the last column (Figure 9.3b). Thus the two Boolean expressions, $\overline{\overline{A}.\overline{B}}$ and $A + B$ are equivalent to each other.

To make a NOR gate, the preceding OR gate must be followed by a NOT gate as shown in Figure 9.4d. This means that a seventh

column is added to the NAND gate truth table, and a fourth
column is added to the truth table of the OR gate so that

$\overline{A.B}$	$\overline{A+B}$
1	1
0	0
0	0
0	0

Since the last two columns are identical, we have to conclude that
the combination of NAND gates in Figure 9.4d performs the same
function as a NOR gate. Moreover, the two Boolean expressions,
$\overline{A.B}$ and $\overline{A+B}$, are equivalent to each other.

It is more difficult to prove that the arrangement of NAND gates
shown in Figure 9.4e is equivalent to an Exclusive-OR gate. But if you
work through the following truth table, you will find the solution very
logical! The last column is the output for a two-input XOR gate since
the output is 1 only when either input, but not both, is at logic 1.
This proves that the combination of NAND gates in Figure 9.4e is
indeed equivalent to an XOR gate (Figure 9.3f).

A B	$P=\overline{A.B}$	$A.\overline{A.B}$	$B.\overline{A.B}$	$Q=\overline{A.\overline{A.B}}$	$R=\overline{B.\overline{A.B}}$	$\overline{(A.\overline{A.B}).(B.\overline{A.B})}$	
0 0	1	0	0	1	1	0	
0 1	1	0	1	1	0	1	(same as XOR
1 0	1	1	0	0	1	1	gate (Figure 9.4c))
1 1	0	0	0	1	1	0	

9.5 Logic families

A logic family is a particular design of logic integrated circuit such
that all members of the family will happily work together in a
circuit design. There are two main families of digital logic.

The older one is TTL (transistor–transistor logic) that was introduced by Texas Instruments Ltd in 1964 based on bipolar transistors. The standard TTL family has been the most popular family of logic ICs ever developed, though one of its subfamilies known as 'low-power Schottky' offers faster speed and lower consumption compared with the original family. All members of the basic TTL family and low-power Schottky operate from a +5 V (± 0.25 V) power supply and are pin-for-pin compatible. That is, the standard '7400' series of TTL logic is compatible with the '74LS00' low-power Schottky series. The first member in the low-power Schottky series is the 74LS00 IC (equivalent to the original 7400 IC), which is a 'quad two-input NAND gate', i.e. it contains four NAND gates each with two inputs and one output. Thus, with two power supply pins, the 74LS00 IC has 14 pins ($4 \times 2 + 4 \times 1 + 2$) and is manufactured in the familiar dual-in-line (d.i.l.) package as shown in Figure 9.5. Some other members of the 74LS00 family are listed in the table below.

Manufacturer's number	What it is
74LS00	quad two-input NAND gate
74LS02	quad two-input NOR gate
74LS04	hex inverter (i.e. six NOT gates)
74LS08	quad two-input AND gate
74LS30	eight-input AND gate
74LS32	quad two-input OR gate
74LS86	quad exclusive-OR gate

If the elements inside a logic IC are made from MOSFETs rather than bipolar transistors, we have the CMOS (or CMOSL) family of logic ICs. CMOS stands for complementary metal-oxide semiconductor logic and refers to the fact that two types of MOSFET are used to make the logic elements, an n-channel MOSFET combined with a p-channel MOSFET (Chapter 8). The best known of the CMOS family is the 4000 series that was introduced in 1968. For example, the 4001 device is designated a 'quad two-input NOR gate'. Like its counterpart the TTL 7400 device, the 4001 is housed in a 14-pin d.i.l. package as shown in

Figure 9.5b. But unlike the TTL family, the 4000 family is capable of operating from any supply voltage in the range 3 V to 15 V, and even to 18 V if the device has a letter 'B' (for 'buffered') after its coding. The 4001 heads the 4000 series, some of which are listed in the table below.

Manufacturer's number	What it is
4001B	quad two-input NOR gate
4011B	quad two-input NAND gate
4069B	hex inverter (i.e. six NOT gates)
4071B	quad two-input OR gate
408lB	quad two-input AND gate
4030B	quad exclusive-OR gate

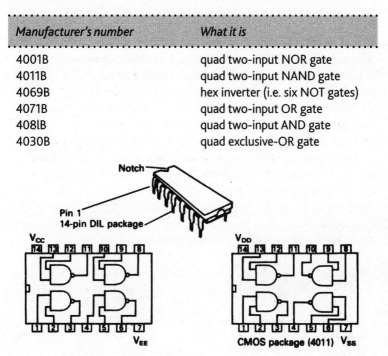

Figure 9.5 TTL/low power Schottky and CMOS versions of the quad two-input NAND gate.

One of the great advantages of the CMOS family is that the devices require less electrical power than TTL devices, although this is now matched by the low-power Schottky series. For this reason, both series of ICs are ideal for portable, battery-operated applications. Another advantage is that each MOSFET transistor that makes up CMOS logic circuits requires only about one fiftieth of the 'floor space' on a silicon chip compared with a bipolar transistor. So MOSFET logic ICs and their variants are ideal for complex silicon chips such as microprocessors and memories – see Chapter 12.

Above all, the great advantage of the CMOS series is that its ICs operate up to a 15 V power supply unlike the TTL and low-power Schottky which need a stabilized 5 V supply.

9.6 Decision-making logic circuits

This section examines some examples to show the way that logic gates are used in decision-making circuits.

Example 1

Draw a logic circuit that satisfies the following conditions: a car can be started only if the gear lever is in neutral, the handbrake is on, and the seat belt is buckled.

First, let us decide that when each condition is satisfied, e.g. the belt is buckled, a logic 1 is input to the logic circuit. (The logic 1 signal could be the closure of, say, a reed switch.)

$$S = A.B.C$$

This Boolean equation summarizes the following function of the logic circuit.

The car can be started (S) if:

► the gear is in neutral (A),
► the handbrake is on (B) AND
► the belt is buckled (C).

The truth table for this Boolean equation contains eight lines as follows:

A	B	C	S = A.B.C
0	0	0	0
0	0	1	0
0	1	0	0
0	1	1	0
1	0	0	0
1	0	1	0
1	1	0	0
1	1	1	1

} car cannot be started

car can be started

A logic circuit that uses two-input AND gates to satisfy this truth table is shown in Figure 9.6. Each AND gate could be replaced by the logic circuit in Figure 9.4b if universal two-input NAND gates were available.

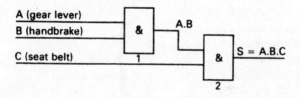

Figure 9.6 Example 1.

Example 2

An alarm system is to protect a room from unauthorized entry through a window, W, and a door, D. The alarm system is armed with a keyswitch, K, and the alarm sounds if either the window or the door is opened. Design a logic circuit that provides the protection required.

Let us assume that the key switch provides a signal at logic 1 when the alarm system is armed, and the opening of the door or the window also provides logic 1.

S = K . (W + D)

This Boolean equation summarizes the following function of the logic circuit.

The alarm sounds if the key switch is armed AND:

► *the door is opened OR*
► *the window is opened OR*
► *the door and window are both opened.*

The truth table for this Boolean equation is as follows:

K	W	D	S = K.(W + D)
0	0	0	0
0	0	1	0
0	1	0	0
0	1	1	0
1	0	0	0
1	0	1	1
1	1	0	1
1	1	1	1

Rows 1–5 (0 0 0 through 1 0 0): } alarm does not sound

Rows 6–8 (1 0 1 through 1 1 1): } alarm sounds

A logic circuit that uses one two-input OR gate and one two-input AND gate is shown in Figure 9.7. If only two-input NAND gates were available, the circuit could be designed by using Figure 9.4c to replace the OR gate and Figure 9.4b to replace the AND gate.

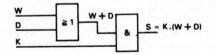

Figure 9.7 Example 2.

Example 3

A chemical plant has four large tanks A, B, C and D that contain different liquids. Liquid level sensors are fitted to tanks A and B and temperature sensors to tanks C and D. Design a logic system that provides a warning when the level in tanks A or B is too high at the same time as the temperature in tanks C or D is too low.

First, let us decide what logic 1 and logic 0 signals represent:

▶ *Logic 1 is a signal that the level in tank A or B is too high, and logic 0 that the level is satisfactory.*
▶ *Logic 1 is a signal that the temperature in tank C or D is too low, and logic 0 that the temperature is satisfactory.*
▶ *A warning signal is represented by logic 1 at the output from the circuit.*

$S = (A \oplus B).(C \oplus D)$

This Boolean equation summarizes the following function of the logic circuit.

A warning is given if:

▶ level in tank A OR tank B is too high AND
▶ the temperature in tank C OR tank D is too low.

The truth table for this Boolean equation, that includes only those lines where S = 1 and a warning is given, is as follows:

A	B	C	D	$A \oplus B$	$C \oplus D$	$S = (A \oplus B).(C \oplus D)$	
1	0	0	1	1	1	1	
1	0	1	0	1	1	1	A warning is
0	1	0	1	1	1	1	given
0	1	1	0	1	1	1	

Note that we are using the exclusive-OR condition; A OR B and C OR D, and not A AND B and C AND D, must be at logic 1 to give an output of logic 1. For example:

A	B	C	D	$A \oplus B$	$C \oplus D$	$S = (A \oplus B).(C \oplus D)$	
1	1	0	0	0	0	0	
0	0	1	1	0	0	0	A warning is not
1	1	1	1	0	0	0	given

A logic circuit that uses two two-input exclusive-OR gates and one two-input AND gate is shown in Figure 9.8. If universal two-input NAND gates were available, each exclusive-OR gate could be replaced by four NAND gates as shown in Figure 9.4e, and the AND gate by Figure 9.4b. However, our decision to use exclusive-OR gates ignores the possibility that the temperature in tanks A and B may be too low simultaneously, and/or the level in tanks C and D may be too high simultaneously. If these conditions are also to provide a warning, can you solve the above problem using NAND gates?

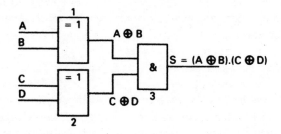

Figure 9.8 Example 3.

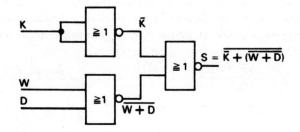

Figure 9.9 An alternative solution for Example 2.

Note that the aim of logic circuit design is to produce a circuit that uses as few universal logic gates as possible. Thus, the use of three two-input NOR gates as shown in Figure 9.9 would provide a solution to Example 2. Draw up a truth table to show that this logic circuit has the same function as the one given on page 172.

Insight

In fuzzy logic, the Boolean logic states of 0 and 1 are simply regarded as special cases of truth, but include the various states of truth in between. For example, instead of comparing two things on the basis of whether they are short or tall, fuzzy logic would reckon on short as being 0.34 of 'tallness'. Surely this is the way our brains come to reasoned judgements about the world around us, in analogue rather than digital ways. We accumulate information to allow us to construct a number of partial truths that we assemble as higher truths so that ultimately, when certain limits are reached, a mental and/or physical response takes place. Fuzzy logic, then, is a superset of conventional Boolean logic that has been extended to handle the concept of partial truth, i.e. values of 'truth' that lie between 'completely true' and 'completely false'. It is hardly surprising that fuzzy logic is applied in computer programs known as 'expert systems' that imitate the judgement and behaviour of a human who has expert knowledge in a particular field. As a rule, an expert system has a knowledge base containing accumulated experience and a set of rules for applying the knowledge base to each particular situation that is described to the program. Sophisticated expert systems can be enhanced with additions to the knowledge base or to the set of rules. Programs that play chess and assist in medical diagnosis are among the best-known expert systems. For more details see: http://www.seattlerobotics.org/Encoder/mar98/fuz/fl_part1.html#INTRODUCTION

TEST YOURSELF

1 *What is the minimum number of inputs that an AND gate can have?*

2 *An EOR gate's output is 1 when .....*

   all inputs are 0 / only one input is 1

3 *Write down the truth table for a NOT gate.*

4 *A NAND gate is called a universal logic gate because:*
   (a) *it turns up everywhere*
   (b) *it can be used to produce all other basic logic functions*
   (c) *it takes a lot of words to describe it.*

5 *Show how to make a NOT gate out of a NAND gate.*

6 *Write down the truth table for a three-input NOR gate.*

7 *Show that the arrangement of two two-input AND gates in Figure 9.10 is equivalent to a single three-input AND gate.*

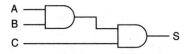

Figure 9.10 Question 7.

8 *Write down the truth table for the logic circuit in Figure 9.11.*

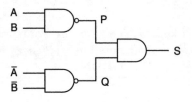

Figure 9.11 Question 8.

9 *What are the main differences between the three logic families 7400 series, 74LS00 series and 4000 series?*

10 *Use truth tables to prove the following relationships are correct.*
   (a) $A.(A + B) = A$
   (b) $A + \overline{A} = 1$
   (c) $(A + B). (A + C) = A + (B.C)$

10

Flip-flops and counters

In this chapter you will learn:
- *to compare the operation of a flip-flop with a digital logic gate*
- *why a flip-flop is a divide-by-two counter*
- *about writing truth tables to show how flip-flops are used in a binary-coded decimal (BCD) counter*
- *about the operation of CMOS frequency dividers*
- *about some uses of binary counters and frequency dividers.*

10.1 What flip-flops do

You have seen that the truth table for a logic gate specifies the binary value (0 or 1) of the gate's output for every possible combination of binary input values. That is why a decision-making circuit using gates is called combinational logic. As well as being used in decision-making circuits, combinational logic circuits are the basic building blocks of encoders and decoders (Chapter 11), and of arithmetic circuits.

Now a flip-flop is also made from logic gates, but its function is quite different from that of combinational logic circuits. A flip-flop 'remembers' its binary data until it is 'told' to forget it. The logic state of its output is determined by the value of the binary data it has stored in its memory, as well as any new data it is receiving. For this reason, circuits built from flip-flops are said to be sequential logic circuits. It is no surprise, then, that flip-flops are the basic memory cells in many types of computer memory (Chapter 12). They are also used in binary counters as explained in this chapter. The flip-flop is also known as a bistable and it is one of a family of multivibrator circuits that includes the monostable and the astable discussed in Chapter 6 in connection with the two configurations of the 555 timer.

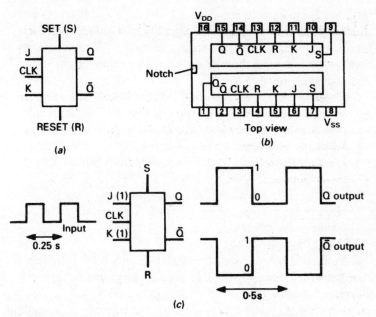

Figure 10.1 The JK flip-flop: (a) its symbol; (b) the 4027 dual JK flip-flop; (c) waveforms produced when a flip-flop toggles.

There are several different types of flip-flop, but the symbol for the most commonly used type is shown in Figure 10.1a. This is a JK flip-flop having two inputs that are called J and K (for no obvious reason). There is one input called the clock (CLK) input that is fed with on/off clock pulses, i.e. a series of 1s and 0s. There are two outputs, Q and $\bar{Q}$, which are complementary, i.e. when output $\bar{Q}$ is at logic 1, output Q is at logic 0, and vice versa. In addition, the flip-flop has a SET input (S) and RESET input (R). A good example of a JK flip-flop in the '4000' series of CMOS digital ICs is the

4027 device. As Figure 10.1b shows, it contains two identical and independent flip-flops.

In this chapter I want to focus on the use of the flip-flop as a toggle flip-flop. This means that when a series of clock pulses are fed into the clock input, each pulse causes the flip-flop outputs to change state from 1 to 0 or vice versa, i.e. to toggle. This behaviour is shown by the waveforms in Figure 10.1c. To make the 4027 toggle, the J and K inputs must equal logic 1 and the S and R inputs must equal logic 0. Now suppose the frequency of the clock pulses is 4 Hz, i.e. the time between consecutive 1s is 0.25 s. This toggle flip-flop has the following three effects on the input waveform.

1 *The frequency of the waveform at each output has been halved by the flip-flop, the output frequency is now 2 Hz so that the time between consecutive 1s and 0s is 0.5 s.*
2 *When the waveform at the $\bar{Q}$ output goes from HIGH to LOW, the waveform at the Q output goes from LOW to HIGH, and vice versa.*
3 *The change in the state of the Q outputs takes place when the clock pulse changes from LOW to HIGH; we say the 4027 is positive-edge triggered.*

The truth table below shows the HIGH (1) and LOW (0) states of the Q and $\bar{Q}$ outputs as the 4027 flip-flop is toggled by the clock pulses.

CLK	Q	$\bar{Q}$
L(0)	H(1)	L(0)
H(1)	L(0)	H(1) a change
L(0)	L(0)	H(1) no change
H(1)	H(1)	L(0) a change
L(0)	H(1)	L(0) no change
H(1)	L(0)	H(1) a change
... and so on		

This table clearly shows that a change in the states of the two outputs occurs when the clock pulse changes from 0 to 1, and a change from 0 to 1 occurs once for every two LOW-to-HIGH changes of the clock pulse. Let us see how useful this behaviour of a toggle flip-flop is in the design of electronic counters.

Insight

In electronic terms a flip-flop is a circuit with two stable states but it has several other meanings. For example, it is also a term used to describe a simple thong sandal; a sudden change of position on some issue especially in politics and a handspring in gymnastics.

10.2 Making binary counters from flip-flops

Suppose that every time a binary 1 'enters' a flip-flop it represents a single count. For example, this count could be the signal produced by a reed switch every time a bicycle wheel rotates (Chapter 3). Thus if 16 counts are input to this flip-flop, eight counts are delivered from its Q output. Figure 10.2 shows what happens if the Q output from this first flip-flop, FF_1, is connected to the clock (CLK) input of a second flip-flop, FF_2. Since each flip-flop divides by two, two flip-flops divide by four $(2 \times 2 = 2^2)$ and four counts leave FF_2. If a third flip-flop, FF_3, is connected to the Q output of FF_2, the overall division is eight $(2 \times 2 \times 2 = 2^3)$ and two counts leave the Q output of FF_3. A fourth flip-flop, FF_4, provides an overall division by sixteen (2^4) and one count leaves FF_4. As shown in Figure 10.2, the connection of four flip-flops is said to be 'cascaded', i.e. one following the other. It is usual to label the four outputs Q_A to Q_D. The pattern of 0s and 1s from these four outputs make up sixteen input counts as shown in the following table. The pattern of 0s and 1s from the Q outputs shows that eight counts leave Q_A for sixteen counts entering flip-flop FF_1. Four counts leave Q_B, two counts leave Q_C, and one count leaves Q_D.

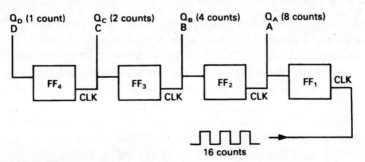

Figure 10.2 A four-bit binary counter using four cascaded flip-flops.

Input count	Q_D(MSB)	Q_C	Q_B	Q_A(LSB)
0	0	0	0	0
1	0	0	0	1
2	0	0	1	0
3	0	0	1	1
4	0	1	0	0
5	0	1	0	1
6	0	1	1	0
7	0	1	1	1
8	1	0	0	0
9	1	0	0	1
10	1	0	1	0
11	1	0	1	1
12	1	1	0	0
13	1	1	0	1
14	1	1	1	0
15	1	1	1	1
16 counts	1 count	2 counts	4 counts	8 counts

The four columns that are identified Q_A to Q_D give the binary equivalent of the number of counts entering the first flip-flop, FF_1. Thus, when eleven decimal counts have entered the first flip-flop, the binary value represented by the 1s and 0s of the flip-flop outputs is 1011. This binary number is equivalent to decimal 11 as can be proved by adding the decimal values of the four binary digits:

$$1 \times 2^0 + 1 \times 2^1 + 0 \times 2^2 + 1 \times 2^3 = 11$$

Since four binary digits make up each number in the table, the counter is called a four-bit binary counter.

The binary digit (or bit, for the sake of brevity) in the Q_A column is known as the least significant bit (LSB) since it has the lowest binary weighting, 2^0, in the number. The bit in the Q_D column is known as the most significant bit (MSB) since it has the highest binary weighting, 2^4. Note that this binary counter counts to decimal 15, from binary 0000 to binary 1111. A five-bit binary counter able to count from 00000 to 11111 (decimal 31) is obtained by adding a fifth flip-flop after FF_4. Six flip-flops would count to binary 111111 (decimal 63), seven to binary 1111111 (decimal 127), and eight to binary 11111111 (decimal 255). It follows that if a binary counter has n flip-flops, there are n bits in the number that can be counted.

Insight

The decimal counting system that we use in everyday life has 10 discrete symbols, the numbers 0 to 9, which are probably a remnant from counting on one's fingers. However, Aborigines who live in the Torres Strait near Australia counting us a base two (binary) system.

A binary-coded decimal (BCD) counter is a version of the four-bit counter that counts to decimal 9 in binary. It is used in many types of digital display, e.g. in digital watches, weighing scales and petrol pumps. Three BCD counters can be connected as shown in Figure 10.3 to display a binary count equivalent to the decimal number 139. Counter Z shows the binary value (1001) of the 'units' digit, counter Y the binary value of the 'tens' digit (0011), and counter X the value (0001) of the 'hundreds' digit. Note that the MSB of each counter is connected to the CLK input of the preceding counter. Every time the MSB changes from 1 to 0, a pulse is passed from counter Z to counter Y, or from counter Y to counter X. Initially all three BCD counters are set to 0000 by applying a reset pulse to all counters simultaneously. After 139 clock pulses, BCD counter Z shows a binary value of 1001, counter Y a binary value of 0011, and counter X a binary value of 0001. How did these three four-bit numbers get there?

Suppose 9 pulses have entered counter Z. Counter Z then shows 1001 and counters Y and X each show 0000. After the next clock pulse, counter Z changes to 0000 and counter Y to 0001. For pulses 11 to 19, counter Y shows a steady reading of 0001 while counter Z passes through the binary counts of 0001 to 1001. After the 20th pulse, counter Y changes to 0010 and counter Z goes back to 0000. The same thing happens to counter Z after 29 pulses; counter Z changes from 1001 to 0000 and counter Y increases to 0011. Thus after 99 pulses, counters Z and Y show 1001 and counter X 0000. After the next clock pulse (decimal 100), both counters Z and Y show 0000, and counter X, 0001. During the next 39 clock pulses counter X shows a steady 0001 while counters Z and Y repeat the changes for the first 39 pulses. The maximum number of clock pulses that can be displayed by these three BCD counters is 999. Chapter 11 explains how the binary numbers from BCD counters such as this one can be displayed in a decimal form on LED and LCD displays.

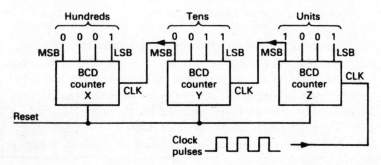

Figure 10.3 Three cascaded BCD counters.

Figure 10.4 The crystal in this watch is the bottle-shaped object below the liquid crystal display.

10.3 Frequency dividers and counters

Computers, digital clocks and watches, and many types of control system depend for their operation on a series of electronic 'ticks' that occur at a particular frequency. For example, mains-operated digital clocks sometimes make use of the 50 Hz mains frequency that is maintained accurately at this value by the electric power stations. Computers, digital watches and clocks use the rapid vibrations of a tiny quartz crystal (Figure 10.4) to provide a stable frequency of, for example, 32.768 kHz. How can flip-flops and binary counters be used to reduce frequencies of 50 Hz and 32 kHz to a frequency of 1 Hz?

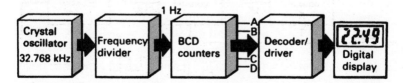

Figure 10.5 The electronic system of a digital watch.

Figure 10.5 shows the system used in a digital watch for reducing the 32.768 kHz generated by a quartz crystal to a frequency of 1 Hz. Here a series of cascaded flip-flops repeatedly divide the frequency by two. The frequency of 32.768 kHz should be divided by two how many times? The answer is fifteen times since two to the power 15 (2^{15}) equals 32,768. These 1 Hz pulses are then delivered to a set of cascaded BCD counters followed by decoders (Chapter 11) so that a digital display can be operated. In a digital watch, the oscillator (except the crystal itself), the flip-flops, BCD counters and decoders are all contained in a single integrated circuit.

The reduction of the mains frequency to 1 Hz requires the system shown in Figure 10.6. First a BCD counter divides the mains frequency by 10 to give 5 Hz. Then a second BCD counter is arranged to divide by 5 giving the 1 Hz frequency required. These 1 Hz pulses are then used to operate BCD counters and decoders to drive the display.

Figure 10.6 The electronic system of a mains-operated clock.

The 4020, 4040 and 4060 devices shown in Figure 10.7 are frequency dividers in the CMOS family of ICs. They contain several cascaded flip-flops to divide by 2^{14}, 2^{12} and 2^{14}, respectively. All the outputs of the internal flip-flops are available from the 4040 device so a maximum frequency division of $2^{12} = 4096$ is possible. Obviously, the 4020 and 4060 devices cannot provide all flip-flop outputs since there are not enough pins on the package! The 4020 skips division by 2^2 and 2^8, and the 4060 skips division by 2^1, 2^2, 2^3 and 2^{11}. The maximum frequency division factor of the 4060 is $2^{14} = 2 \times 2 \times 4096 = 16,384$. Thus the 4040 could be used to divide the crystal frequency in Figure 10.5 by 16,384 to provide an output frequency of 32,768/16,384 = 2 Hz; one further flip-flop would provide 'ticks' at intervals of one second.

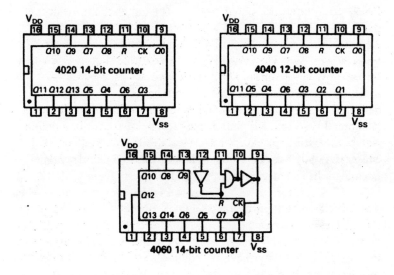

Figure 10.7 Examples of CMOS frequency dividers.

These frequency dividers are easy to use. In normal use, the counter counts every time the input signal changes from HIGH to LOW. For counting to take place, the RESET input (e.g. pin 11 on the 4040) must be connected LOW, to 0 V. If the RESET is taken HIGH, the counters reset to zero, so all outputs become logic 0. The 4510 and 4516 CMOS devices shown in Figure 10.8 are both four-bit binary counters, the former a BCD counter with a maximum binary count of 1001 (decimal 9), and the latter a four-bit counter that has a maximum count of 1111 (decimal 15). Both devices contain four cascaded flip-flops so they could replace the design based on individual flip-flops shown in Figure 10.3. And both devices can count up or down in binary depending on whether the UP/DOWN terminal (pin 10) is set at logic 1 or logic 0, respectively. Individual four-bit counters can be cascaded by connecting the CARRY OUT (CO) pin of one device to the CARRY IN (CI) pin of the next device. When used as a binary counter, the CARRY IN, RESET and LOAD terminals are set at logic 0. There are other connections on these ICs but these need not worry us here.

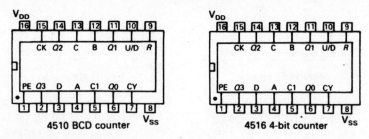

Figure 10.8 Examples of CMOS four-bit and BCD counters.

The CMOS 4017 device shown in Figure 10.9 is a decade counter, meaning it has ten decimal outputs and an overflow, or CARRY OUT (CY – just for CARRY), output. As the waveforms show, when its RESET terminal (pin 15) and CLOCK ENABLE (pin 13) terminal are at logic 0, these ten outputs go HIGH each time the clock pulse goes from LOW to HIGH. At any point in the counting sequence, all the LEDs except LED$_0$ (the 'zero' count on pin 3) can be switched off by connecting the RESET pin to logic 1, momentarily.

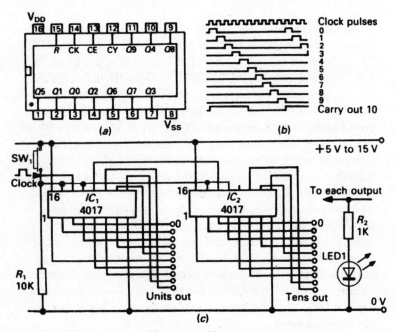

Figure 10.9 The CMOS 4017 decade counter: (a) its appearance, (b) output waveforms.

188

10.4 Binary adders

Calculators and computers manipulate binary numbers using circuits that add and subtract. At the heart of a computer's microprocessor is an arithmetic and logic unit (ALU) that performs arithmetic on binary numbers. The most basic function of the ALU is called the half-adder. The half-adder does what we do mentally when we add two binary digits; $0 + 0 = 0$, $1 + 0 = 1$, $1 + 1 = 10$. Figure 10.10 shows one way of designing a circuit to add two single-bit binary numbers. It is known as a half-adder and is made from an exclusive-OR gate and an AND gate. Since the circuit has two inputs, there are 2^2, four, combinations of inputs to consider and they are summarized in the following table.

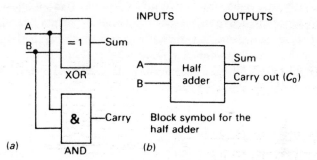

Figure 10.10 A half adder: (a) logic diagram; (b) circuit symbol.

A B	sum $= A + B$	Carry $= A \cdot B$
0 0	0	0
0 1	1	0
1 0	1	0
1 1	0	1
bits added	XOR gate	AND gate

The 'sum' column shows why an exclusive-OR gate is necessary, while the 'carry' column justifies the use of an AND gate. Thus, the half-adder can only deal with the addition of the least significant bit. Clearly we need a system that not only adds A and B but is also able to cope with the carry bit. Such a system is called a full-adder. Figure 10.11 shows one way of constructing it from two half-adders and one OR gate. It accepts two binary digits, A and B, plus a carry-in bit, C_{in}. The eight lines in the truth table below show all the additions the full-adder can do.

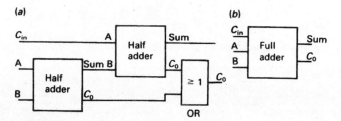

Figure 10.11 A full adder: (a) logic diagram; (b) circuit symbol.

inputs			outputs	
carry-in, C_{in},	B	A	sum	carry-out, C_{out}
0	0	0	0	0
0	0	1	1	0
0	1	0	1	0

inputs			outputs	
carry-in, C_{in}	B	A	sum	carry-out, C_{out}
0	1	1	0	1
1	0	0	1	0
1	0	1	0	1
1	1	0	0	1
1	1	1	1	1

carry + A + B

For instance, suppose A = 1, B = 1 and C_{in} = 0 as shown in Figure 10.12a. The first half-adder has a sum of 0 and a carry of 1. The second half-adder has a sum of 0 and a carry of 0. Thus the final output is a sum of 0 and a carry of 1 as summarized in line 4 of the truth table. If A = 1, B = 1, and C_{in} = 1 as shown in Figure 10.12b, the final line of the truth table indicates that we get a sum of 1 and a carry of 1. Two four-bit binary numbers can be added together using three full-adders and one half-adder to produce a parallel binary adder. IC manufacturers produce several types of adders. The low-power Schottky 74LS181 and the CMOS 4581 each have 24 pins and are both capable of adding two four-bit words, $A_3A_2A_1A_0$ and $B_3B_2B_1B_0$ producing the sum at their four outputs, $Q_3Q_2 Q_1Q_0$.

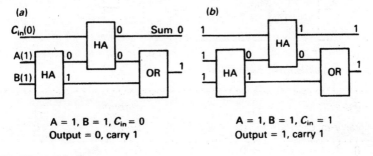

Figure 10.12 Using full-adders.

TEST YOURSELF

1 The two outputs of a flip-flop are known as complementary outputs because:
   (a) when in use they warm up
   (b) they always have opposite logic states
   (c) they act in unison.

2 Draw a diagram to show how two flip-flops can be connected to produce a two-bit counter.

3 Explain with a truth table the function of a four-bit BCD counter.

4 If a signal of frequency 512 kHz is input to seven cascaded (connected one after the other) flip-flops, what is the frequency of the signal that emerges from the seventh flip-flop?

5 Four BCD counters are cascaded.
   (a) What is the maximum decimal count possible?
   (b) If the BCD counters are reset and 369 counts are delivered to the 'units' counter, what do the BCD counters read?

6 Draw a diagram to show how binary counters can be used to give a 1 Hz pulse from the 50 Hz mains supply.

7 Five cascaded BCD counters are set to zero, and a gating pulse exactly 2 s wide allows pulses at a frequency of 32 kHz to enter the counters. What are the outputs of the five counters at the end of the gating pulse?

8 Explain the statement that 'a flop-flop is a divide-by-two counter'.

9 Explain the function of a 4017 decade counter.

10 *Figure 10.13 shows a four-bit binary counter that is fed with a regular series of clock pulses. The AND gate selects two of the four bits from the four-bit counter and its output is fed to the reset terminal of the counter. Explain why the counter counts 10 pulses before resetting to zero.*

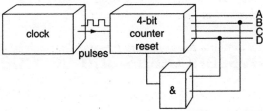

Figure 10.13 Question 10.

11

....

Displays, encoders and decoders

In this chapter you will learn:

- *the difference between analogue and digital displays*
- *the way numbers and characters are created by a seven-segment display*
- *the difference between a light-emitting diode (LED) display and a liquid crystal display (LCD)*
- *about decoders and encoders*
- *the function of a BCD-to-seven-segment decoder/driver*
- *the principle and function of multiplexing LED displays.*

11.1 Analogue and digital displays

You have only to think of the array of instruments in the cockpit of a modern airliner, or the control room of a power station, or the advertising displays at a football match to realize that the most convenient way of conveying information is to use some form of visual display. The displays used are generally of two types, analogue or digital, but sometimes a combination of both types. These terms were used in Chapter 4 to describe analogue and digital multimeters. An analogue multimeter displays the value of a measurement on a moving-coil meter that uses a pointer moving over a calibrated scale. It registers all possible values between maximum and minimum. On the other hand, a digital multimeter generally uses a liquid crystal display (LCD) or a seven-segment light-emitting diode (LED) display to give a numerical value of a measurement. It registers measurements in discrete steps as the numbers change.

Digital displays based on LCDs and LEDs have largely replaced analogue displays in many different types of instrument. The main advantage of digital displays is that, having no moving parts, they are more rugged and can stand up to vibration better than the rather fragile moving-coil meter. They are also cheaper and easier to manufacture, and purpose-designed ICs are readily available to operate them. But perhaps the main reason for their rise to fame is that many of today's electronic systems process digital signals that are compatible with the operating principles of LCDs and LEDs.

But remember, a numerical display is not always the preferred choice in a digital system. Sometimes it is better to use an analogue display when the change in a reading is looked for. Analogue displays are often used on hi-fi amplifiers in preference to digital displays for indicating the audio power delivered to loudspeakers or the signal strength of a radio station. These analogue displays use a 'bar of light' made of discrete LEDs or LCD segments that lengthens or shortens in response to the signal strength. Analogue displays of this kind make it easier to see how the signal strength changes with time rather than having to interpret the precise value given by a numbered display. Perhaps that is why some people prefer digital watches with LCD 'hands' since the time of day seems to have more meaning when set against the twelve-hour scale of time round the face of the watch.

11.2 The seven-segment LED display

The combination of electronics and optics to display information is known as optoelectronics. For example, an LED (Chapter 7) is an optoelectronics device. Numbers, letters and other symbols are formed by the selective illumination of one or more segments arranged in the form of the figure '8'. Each of the LEDs labelled 'a' to 'g' can be switched on or off by digital circuits. A display of this type, forming both numbers and some letters, is known as an alphanumeric display. Figure 11.1b shows how to display the decimal number 6 by switching on segments 'c' to 'g'. The decimal number 1 is displayed by switching on segments 'b' and 'c', and so on. Only the decimal numbers 0 to 9, a few special symbols such as '–', and a few letters such as 'C', 'c', and 'F' can be displayed by these seven-segment displays. A decimal point can be displayed by illuminating an eighth LED normally placed to the right of the digit.

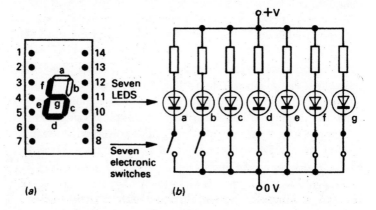

Figure 11.1 The seven-segment LED display: (a) how the segments are labelled; (b) how decimal '6' is obtained.

The digital circuits needed to light particular segments of the seven-segment display are described below. There are two types of seven-segment display depending on the nature of the digital circuits used. In the common-anode type shown in Figure 11.2a, the anodes of all seven LEDs are connected to the positive terminal of the power supply. To light one of the segments, the cathode terminal of the segment LED is provided with a LOW signal by the digital circuits. In the common-cathode type shown in Figure 11.2b, the cathodes of all seven LEDs are connected to the 0 V terminal of the power supply. To light one of the segments, the anode terminal of the segment LED is provided with a HIGH signal from the digital circuits.

As explained in Chapter 7, the LEDs are made from specially doped semiconductors, mainly gallium arsenide. The colour of the light emitted from the LED depends on the type of 'impurity' introduced into the crystal structure of gallium arsenide. In this way, LEDs that emit red, green, yellow, blue and even white light can be manufactured.

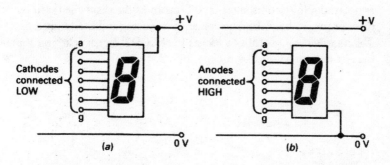

Figure 11.2 (a) A common-anode seven-segment LED display requires a LOW drive signal, and (b) a common-cathode seven-segment LED display requires a HIGH drive signal.

11.3 The liquid crystal display (LCD)

The LCD is a popular method of displaying information, especially in digital watches and pocket games. LCDs can display not just numerical data, but also words and pictures. Large-area LCDs rather than a cathode-ray tube are commonly used on some oscilloscopes and on laptop computers and scientific calculators. The main reason for choosing LCDs for these applications is that their power consumption is minute compared with LED displays. Whereas the LED display requires electrical power to generate light, the LCD simply controls available light. This means that it is easily seen in bright sunlight although it cannot be seen in the dark unless the display is 'backlighted'.

The LCD relies on the transmission or absorption of light by certain organic carbon crystals that behave as if they were both solid and liquid, that is, their molecules readily take up a pattern as in a crystal and yet flow as a liquid. In the construction of the common LCD unit shown in Figure 11.3, this compound is sandwiched between two closely-spaced, transparent metal electrodes that are in a pattern, e.g. as a seven-segment digit. When an a.c. signal is applied across a selected segment, the electric field causes the molecular arrangement of the crystal to change, and the segment shows up as a dark area against a silvery background. A polarizing filter on the top and bottom of the display enhances the contrast of black against silver by reducing reflected light. This type of LCD is called a field-effect LCD, since it relies on the electric field produced by the a.c. signal.

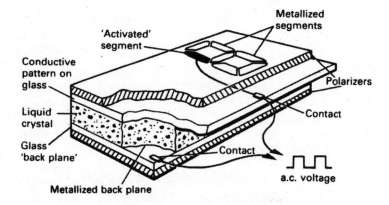

Figure 11.3 The structure of a liquid crystal display.

Insight

Today many people hover between two worlds, the analogue world and the digital world, sometimes crossing over into the digital world but always returning to the familiar analogue world in which we are all embedded. In the analogue world our senses experience qualities such as touch, taste, smell, pressure and temperature that change smoothly from one value to the next. On the other hand, the digital world seems to be a rather desolate world of binary numbers. Yet the austere computer provides exciting and creative opportunities for anyone willing to acknowledge its power to enlarge our understanding of the analogue world that is in our care; for example, our understanding of the impact of global warming is dependent on the modelling and predictive power of the computer – in the hands of talented programmers of course! Indeed, we need to remember how deeply immersed we are in the real analogue world and that the computer is simply a tool (albeit a powerful one) that we harness in order to explore our analogue world more deeply.

11.4 Decoders and encoders

In digital electronics, decoders and encoders are important output and input functions of a system. They are code translators, i.e. they change information from one form to another. These two different functions are shown in Figure 11.4:

▶ *a decoder is used to alter the format of information taken from a system, and*
▶ *an encoder is used to alter the format of information to allow it to enter into a system.*

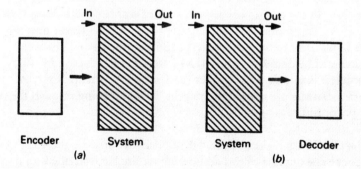

Figure 11.4 In an electronic system, (a) encoders are input devices, and (b) decoders are output devices.

An example of a decoder is a BCD-to-seven-segment decoder-**driver**. This device is required to change the four-bit code produced by a BCD counter into signals capable of displaying decimal numbers on a seven-segment display. An example of an encoder is a telephone keypad that converts a telephone number into a string of binary digits for transmission along a telephone line. Decoders and encoders are examples of combinational logic devices, and it is quite possible to construct them from AND, XOR, etc. logic gates. However, purpose-designed integrated circuit packages are available to make the job easier for the circuit designer. For example, decoders are used in seven-segment displays for converting four-bit BCD codes into patterns of 1s and 0s for operating the LEDs in seven-segment LED displays. These decoders are known as seven-segment decoder/drivers. They are called decoder-drivers since they decode from binary to decimal and provide the necessary current to light, or 'drive', the LEDs in the display.

As shown in Figure 11.2, a seven-segment LED display has its seven segments identified by the letters 'a' to 'g'. In the low-power Schottky digital devices, the 74LS47 device is a decoder-driver designed to drive the segments on or off so as to display the decimal numbers 0 to 9. The connections required between the 74LS47 decoder/driver and the segments of a common-anode display are shown in Figure 11.5. The output terminals of the 74LS47 decoder/driver are connected to the corresponding cathode terminals of the segment LEDs. Note that in this common-anode display, all the anodes of the segment LEDs are connected together to +5 V. This means that the outputs a to g of the 74LS47 have to go LOW, 0 V, to light the corresponding segments and produce a number. The outputs of the 74LS47 are said to be 'active LOW' so that it is said to 'sink' (rather than 'source') current. Note that series resistors are needed to limit the current flowing through the segment LEDs.

The CMOS equivalent of the 74LS47 decoder/driver is the 4511 device. The outputs of this decoder/driver are 'active HIGH' so it has to be used with a common-cathode display. Each output of the 4511 'sources' up to 5 mA of current when operated from a 10 V power supply. The 4511 decodes the ten four-bit binary numbers, 0000 to 1001, and produces the following patterns of 0s and 1s for driving the seven segments of a common-cathode display.

Note that the 4511 does not activate segment 'd' when showing decimal 9, or segment 'a' when showing decimal 6.

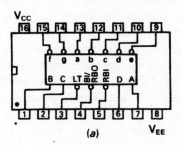

(a)

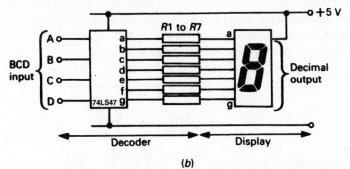

(b)

Figure 11.5 The 74LS47 decoder/driver: (a) its pin connections, and (b) the connections to a common-anode seven-segment LED display.

BCD inputs				Segment outputs							Number displayed
D	C	B	A	a	b	c	d	e	f	g	
0	0	0	0	1	1	1	1	1	1	0	0
0	0	0	1	0	1	1	0	0	0	0	1
0	0	1	0	1	1	0	1	1	0	1	2
0	0	1	1	1	1	1	1	0	0	1	3
0	1	0	0	0	1	1	0	0	1	1	4
0	1	0	1	1	0	1	1	0	1	1	5
0	1	1	0	0	0	1	1	1	1	1	6
0	1	1	1	1	1	1	0	0	0	0	7
1	0	0	0	1	1	1	1	1	1	1	8
1	0	0	1	1	1	1	0	0	1	1	9

In a liquid crystal display (LCD) there must be no d.c. signal across the display. So, a special form of drive circuit is needed as shown in Figure 11.6. The BCD decoder is receiving the binary number 1001 (decimal 9). The decoder therefore activates the 'a', 'b', 'c', 'f' and 'g' outputs since the decoder outputs are active HIGH; and the 'd' and 'e' outputs are LOW. The backplane of the display receives a 30 Hz square wave signal that is also applied to each of the CMOS Exclusive-OR gates used to drive the LCD. The XOR gates ensure that when the segment input is HIGH, the segment drive voltage and the backplane voltage are exactly 180° out of phase, they are the inverse of each other. Thus, there is an overall a.c. voltage across the liquid crystal, resulting in this segment being dark. When the segment input is LOW, the segment input and the backplane are exactly in phase. There is no longer a voltage across the liquid crystal and the segment remains transparent.

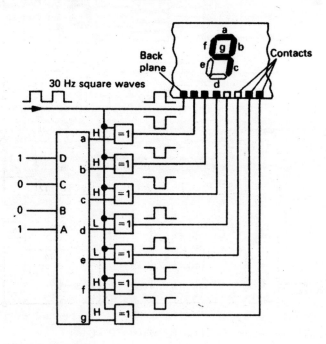

Figure 11.6 The drive circuit required for an LCD.

11.5 Multiplexing seven-segment displays

Each digit in a digital display requires one BCD counter and one decoder/driver for a two-digit display based on TTL devices as shown in Figure 11.7. In this circuit, when the 'units' digit changes from 9 to 0, the 'tens' digit increases by 1. In common with digital watches, clocks and other displays, this simple display has 'leading zero blanking' (LZB). This means that for counts below 10 the 'tens' digit is blanked, i.e. it does not show up as '0'. LZB saves on battery power and makes the display easier to read.

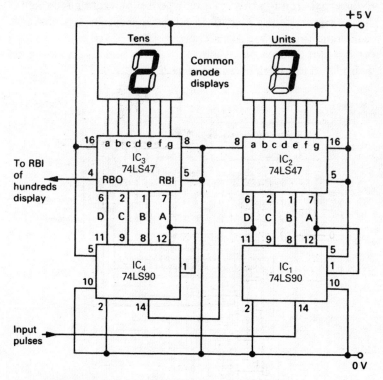

Figure 11.7 Using low-power Schottky devices in a two-digit counter.

If we wanted to extend the simple two-digit display to three, four or more digits, the circuit becomes complicated since each digit requires its own decoder/driver. In addition, all the digits are switched on at the same time, which is quite a drain on battery power. However, there is a technique that reduces circuit complexity by using just one decoder/driver, and which reduces the drain on the power supply by switching the digits on one at a time in rapid succession. The technique is called multiplexing.

Multiplexing scans the digits one after the other so rapidly that they all appear to be on due to the eye's persistence of vision. A flashing light that occurs at a frequency above 20 to 30 Hz appears to the eye to 'run together' to give the effect of a continuous light. A multiplexed display illuminates all the appropriate segments of each digit for a fraction of a second, and immediately after does the same for the next one. Figure 11.8 shows how this is done for a multiplexed four-digit seven-segment display.

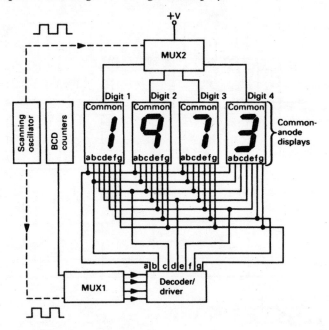

Figure 11.8 Multiplexing a multidigit display.

The scanning oscillator synchronizes the two multiplexers, MUX1 and MUX2. These act like single-pole four-way rotary switches. MUX1 takes the four outputs from the four BCD counters and supplies them as inputs to the BCD decoder/driver. At the instant that the BCD signal is routed to a particular digit, MUX2 energizes that digit by connecting the cathodes of its seven LEDs to 0 V (assuming they are common-anode devices). By varying the period during which each display is off, the brightness of the display can be varied. This is usually achieved by varying the mark-to-space ratio of the clock operating the multiplexers. Since each segment LED is only illuminated for a short time, fairly high maximum currents can be passed through each segment, resulting in a bright display, yet one that requires, on average, less current than a non-multiplexed one. All the circuitry – counters, decoders, multiplexers, etc. – required to drive four or more digits in digital displays is readily available in integrated circuit packages.

TEST YOURSELF

1 Compare the advantages and disadvantages of analogue and digital watches with respect to ease of use and other factors.

2 What is the function of
   (a) an encoder
   (b) a decoder in electronic systems?

3 What are the characteristics of an alphanumeric display?

4 What is leading zero blanking in connection with seven-segment displays?

5 State two advantages of a liquid crystal display compared with a light-emitting diode display.

6 What is the value of the resistor that you must connect in series with an LED to limit the current through it to 10 mA? Assume the voltage across the LED is 2 V and it is operated from a 12 V supply.

7 Draw a seven-segment display and label the segments. What numbers and letters can you obtain from this display?

8 What does it mean if a seven-segment LED display is 'multiplexed'? Why is this done?

9 Describe three applications of LED matrix displays.

10 Give three examples where a single LED is used as a status indicator.

12

Memories and silicon chips

In this chapter you will learn:

- *the purpose of memory in digital systems*
- *about different types of electronic memory including the structure and use of RAM and ROM*
- *the properties and purposes of flash memory and the memory stick*
- *about Moore's Law*
- *about the packaging of integrated circuits*
- *the main steps in the making of a silicon chip.*

12.1 Introduction

Just like human memory, electronic memory is simply any device that stores information for future use. In an electronic memory, this information is stored as a collection of binary digits (or bits), i.e. 1s and 0s. Memory is therefore a feature of digital systems and not of analogue systems. The information, or data, is usually stored in digital memory as groups of 8-bit words (known as bytes), 16-bit or 32-bit words. Computers, calculators, electronic games, hand-held GPS (global positioning systems), digital cameras, mobiles and an increasing number of other digital systems make extensive use of electronic memory.

The way that memory is used in computers is shown in the block diagram of Figure 12.1. Here data flows between the memory devices and a central processor unit or CPU (also called a microprocessor), and between input and output devices. The CPU is the portion of a computer system that carries out the instructions of a computer program. In a computer system such as this, memory is organized either as random-access memory (RAM), or as read-only memory (ROM). Both types of memory are made from thousands of transistors as part of an integrated circuit on a small chip of silicon.

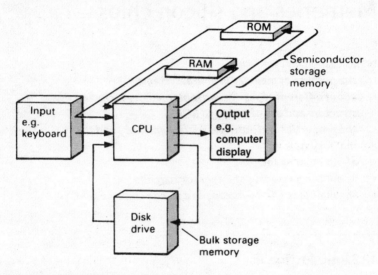

Figure 12.1 The basic building blocks of a computer system.

12.2 Random-access memory (RAM)

RAM is the place in a computer where the computer's operating system, application programs, and data in current use are kept so that they can be reached quickly by the computer's central processor unit. RAM is much faster to read from and write to than the other kinds of storage in a computer, e.g. the hard disk, floppy disk and CD-ROM. However, the data in RAM stays there only as long as your computer is running but that data is lost when you turn the computer off and they are thus known as 'volatile' memory devices. When you turn your computer on again, your operating system and other files are once again loaded into RAM, usually from the computer's hard disk. Some people prefer to call the RAM a 'readily alterable memory' since it is a device whose contents can be altered quickly and easily.

RAM is usually described in terms of the number of bytes of data that it can store. For example, the laptop I am using to write this has 2 gigabytes (2×10^9) of data while its hard disk can hold 200 gigabytes (200×10^9) of data. RAM is situated in a computer as discrete microchips that plug into sockets on the computer's motherboard. On the other hand the hard disk stores data on its magnetized surface. If you have a lot of RAM in your computer, it reduces the number of times that the computer's microprocessor has to read data in from your hard disk, an operation that takes much longer than reading data from RAM. So the more RAM you have in your computer, the faster a program runs and this is especially significant for computer games.

RAM is called 'random access' because any storage location can be accessed directly by reading an address where the data is stored. The term random was originally used to distinguish RAM from storage that is sequential, such as magnetic tape, in which data can only be accessed starting from the beginning of the tape and finding an address sequentially. Thus, it might have been better if RAM had been called 'non-sequential' rather than random since the process of accessing data from RAM is anything but random! Note that other forms of storage such as the hard disk and CD-ROM are also accessed directly (or 'randomly') but the term 'random' access is not applied to these forms of storage.

RAM can be divided into (a) main RAM, which stores every kind of data and makes it quickly accessible to a microprocessor and (b) video RAM, which stores the data or your display screen, enabling images to get to the computer's display faster. The following description concentrates just on main RAM of which there are two types: static RAM (or SRAM) and dynamic RAM (or DRAM).

12.3 Static and dynamic RAM

The essential difference between static and dynamic RAM is the way in which bits are stored in the RAM chips. In a static RAM, the bits of data are written in the RAM just once and then left until the data is read or changed. In a dynamic RAM, the bits of data are repeatedly rewritten in the RAM to ensure that the data is not forgotten. Static RAM is more expensive than dynamic RAM, but, unlike dynamic RAM, does not need to be power-refreshed and the data stored in it is therefore faster to access. Flip-flops (Chapter 10) are the basic memory cells in static RAM. Each flip-flop is based on either two bipolar transistors, or on two metal-oxide semiconductor field-effect transistors (MOSFETs). It is necessary to have as many of these memory cells as there are bits, i.e. 1s and 0s, to be stored.

Dynamic random access memory (DRAM) is the most common kind of random access memory for personal computers and workstations. However, unlike a static RAM (that holds its data until 'told' to change it), a dynamic RAM continually needs to have its storage cells refreshed, or given a new electronic charge every few milliseconds. DRAMs are based on metal-oxide semiconductor field-effect transistors (MOSFETs – see Chapter 8). Bits of data are stored in DRAMs as small packets of charge, rather than as current flow as in the static RAM. This has the advantage that the power consumption of DRAMs based on MOSFET memory circuits is very low.

As shown in Figure 12.2, each memory cell in a DRAM is a very simple circuit and comprises a small capacitor and a single n-channel MOSFET that is switched on to read this charge. Each cell holds a 1 bit as a tiny electrical charge of about 10^{-15} coulombs. Though tiny, this charge still amounts to about 5000 electrons! However, the charge on the capacitor tends to leak away and extensive 'refresh' circuitry is needed to keep the charge 'topped up'. The additional electronics required to ensure that a dynamic RAM's memory does not forget what is in it adds to the cost and complexity of dynamic RAMs. However, the newer dynamic RAMs have refresh circuitry on the chip with the memory cells.

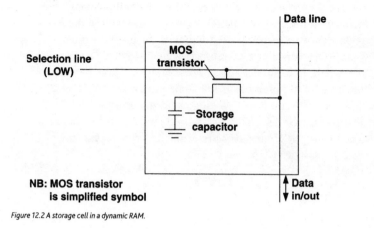

Figure 12.2 A storage cell in a dynamic RAM.

12.4 Read-only memory (ROM)

As mentioned above, the problem with a random-access memory is that its memory is volatile, it loses all its data when the power supply is switched off. A non-volatile memory is a permanent memory that never forgets its data. One type of non-volatile memory is the read-only memory (ROM). ROM is 'built-in' computer memory containing data that normally can only be read, not written to, which is why it is called a 'read-only' memory. Every computer comes with a small amount of ROM that holds just enough programming so that the operating system can be loaded from disk into RAM each time the computer is turned on. Unlike a computer's random access memory, the data in ROM is not lost when the computer power is turned off.

12.5 Flash memory

Flash memory (sometimes called flash RAM) is a type of non-volatile memory that can be erased and reprogrammed in units of memory called blocks. Flash memory is often used to hold control code such as the basic input/output system in a personal computer. When the basic input/output system needs to be changed (rewritten), the flash memory can be written to in block (rather than byte) sizes, making it easy to update. On the other hand, flash memory is not as useful as random access memory because RAM needs to be addressable at the byte not the block level.

The name 'flash memory' arises because the microchip is organized so that a section of memory cells is erased in a single action or 'flash'. The erasure principle is caused by what is called 'Fowler–Nordheim tunnelling' in which electrons pierce through a thin dielectric material to remove an electronic charge from a 'floating gate' associated with each memory cell. Flash memory is used in mobile phones, digital cameras, notebook computers, and other devices. Figure 12.3 shows one type of flash memory used in a digital camera for image recording.

Figure 12.3 The appearance of one form of flash memory used in a digital camera.

A particularly useful form of flash memory is the memory stick. This is a memory module that is so slim and lightweight that it can be carried on a key ring. It has a standard USB plug that can be plugged into the USB port of a computer where it is easily recognized as an external drive. The memory stick was launched by Sony in 1998 and is commonly used to share and transfer pictures, sound and other data between different electronic devices such as laptops, digital cameras and camcorders, indeed any device that has a USB port. One form of memory stick is shown in Figure 12.4. Memory sticks are available in capacities ranging from 128 MB to 16 GB or more. The memory stick family includes Memory Stick PRO that enables a greater maximum storage capacity and faster transfer speeds. Memory Stick Duo is a smaller version of the memory stick and there is an even smaller Memory Stick Micro. In 2006 Sony introduced Memory Stick PRO-HG which is a high speed variant of the PRO intended for use in high definition still and video cameras.

Figure 12.4 A memory stick.

12.6 Gallium arsenide

Gallium arsenide (GaAs) is a crystalline substance that is rival to silicon for making diodes, transistors and integrated circuits. Its main claim to fame is that semiconductor devices made from it conduct electricity five times faster than those made from silicon. This makes GaAs components useful where information needs processing at ultra-high radio frequencies, and in fast electronic switching applications. Needless to say, this property makes gallium arsenide an interesting material for weapons manufacturers since data in computer memories made from GaAs can be accessed more quickly than memories made from silicon. This means that missiles and other weapons can respond rapidly to information received by their sensing and control circuits.

However, there are a few drawbacks to the use of GaAs, one of which is that the two elements gallium and arsenic from which it is made are in short supply, mainly being found as impurities in aluminium and copper ores. On the other hand, silicon is plentiful, being found in silicates such as sand. The cost of GaAs is about 30 times that of silicon, and this is aggravated by the fact that about 90 per cent of GaAs chips are rejected after production. Furthermore, it is not as easy to make integrated circuits from

GaAs as it is from silicon since it does not form a protective layer of oxide to resist the diffusion of dopants during the process of photolithography.

It is therefore unlikely that there will be a rapid rise in the use of GaAs-based semiconductors in the near future except for specialist applications where cost is not a major consideration, e.g. ballistic missile development and the USA's 'Star Wars' programme, the Strategic Defence Initiative (SDI). However, manufacturers of GaAs have turned to space for help in overcoming the problems and costs of producing pure GaAs. A number of countries are planning to produce pure crystalline GaAs in the extremely low vacuum and zero gravity in laboratories aboard orbiting space stations.

12.7 Moore's law

An integrated circuit comprises microscopically small transistors, diodes, resistors and other components connected together on a 2 to 5 mm square chip of silicon. The circuit produced has all the components necessary for the chip to function as an amplifier, or an analogue-to-digital converter, or a memory, or a microprocessor, for example. Integrated circuits contain components that are the smallest objects created by humans; and the trend is towards producing even smaller components on a silicon chip in order to provide more processing power in a smaller space.

In the 1960s, this trend towards increasing complexity was noticed by Gordon Moore, cofounder of the Intel Corporation in the USA. He observed an exponential growth in the number of transistors per integrated circuit and predicted that this trend would continue (Figure 12.5). The doubling of the number of transistors every couple of years became known as Moore's law and still holds true today and it is astonishing to think that it is possible to integrate 42 million devices on a Pentium 4 microprocessor chip just a few millimetres square. Forty-two million may sound a lot but in 2009 Intel produced a central processor unit containing two billion transistors.

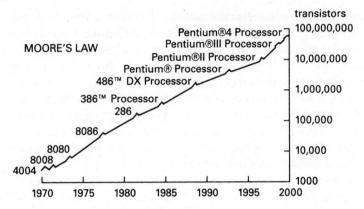

Figure 12.5 Graph showing progressive increase in chip complexity.

12.8 Making a silicon chip

The process of putting several hundred thousand, let alone two billion, transistors on a silicon chip a few millimetres square is not an easy one. A technique called photolithography is at the heart of the process. Photolithography uses photographic techniques and chemicals to etch a minutely detailed pattern on the surface of a silicon chip. Each stage in the process involves the use of photomasks that are prepared photographically. Each photomask holds a particular pattern identifying individual transistors, conducting pathways, etc. Photomasks are produced by the photographic reduction of a much larger pattern. A photomask is placed over a thin layer of photoresist covering the surface of the silicon. When ultraviolet light is shone on the photomask, it passes through the clear areas but the opaque areas stop it. According to the type of photoresist used, either the exposed or the unexposed photoresist can be dissolved away using chemicals to leave a pattern of lines and holes. This pattern enables transistors to be formed in the silicon, and aluminium interconnections to be made between them.

The process of making a silicon chip begins with a 50 to 30 mm or more cylinder-shaped single crystal of pure silicon (or gallium arsenide) known as a boule or ingot (Figure 12.6). The ingot is obtained by slowly pulling the growing crystal from a bath of pure molten silicon. It is then cut up into thin slices known as wafers, about the size of a beer mat and half as thick. The wafers are then passed through an oven containing gases heated to about 1200°C. The gases diffuse into each wafer to give it the properties of a p-type or an n-type semiconductor (Chapter 2) depending on the gas used, a process known as epitaxial growth. The wafers are then ready to have integrated circuits formed on them by a complex process that involves masking, etching and diffusion.

Figure 12.6 A single crystal of pure gallium arsenide and some of the semiconductor products made from it.
COURTESY: DIVISION FOR INTERNATIONAL SCIENCE RELATIONS, TIME

Figure 12.7 shows the several stages required to produce openings in the surface of the silicon through which gases are diffused to create transistors. The first stage involves heating the wafer to about 1000°C in a stream of oxygen so that a thin layer of silicon dioxide is formed over the whole surface of the wafer.

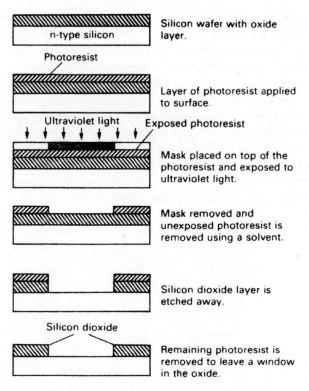

Figure 12.7 Steps in the formation of a window in the silicon dioxide surface of a silicon chip.

In the next stage, a thin layer of a light-sensitive emulsion (photoresist) is spread over the layer of silicon dioxide. A photographic plate (the photomask) is placed over the top of the emulsion. The photomask contains a pattern of dots in microscopic detail that are to become holes in the silicon dioxide layer. A single mask holds the pattern for many hundreds of integrated circuits for each wafer.

In the third stage, the mask is exposed to ultraviolet light. Where the mask is transparent, the light passes through and chemically changes the photoresist underneath so that it hardens. The unexposed photoresist can be removed easily with a suitable solvent (fourth stage).

In the fifth stage, the silicon wafer is immersed in another solvent that removes the silicon dioxide from the unexposed areas. The wafer now has a thin surface layer of silicon dioxide in which there are a large number of tiny 'windows'. It is through these windows that gases are allowed to pass into the epitaxial silicon layer underneath to form transistors. In the production of complete silicon chips on a wafer, the formation of a layer of silicon dioxide, followed by masking and etching, has to be repeated many times.

Figure 12.8 shows the various steps required to make an npn transistor on a silicon chip, and Figure 12.9 shows that similar steps are required to create a single n-channel MOSFET.

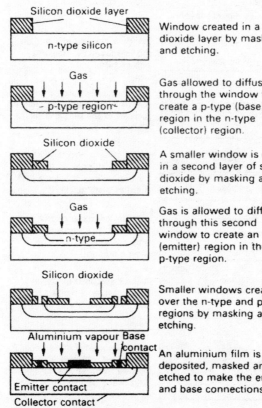

Figure 12.8 Steps in the formation of a single npn transistor on a silicon chip.

First a gas is selected that diffuses through a window to form a p-type base region in the n-type silicon epitaxial layer. Next, a fresh silicon dioxide layer is formed over the window, followed by a stage of masking and etching to create a second smaller window. Through this window, a gas diffuses to form the n-type emitter region. Another layer of silicon dioxide is formed over this window followed by masking and etching to create even smaller windows for making contacts to the base and emitter regions. These contacts are formed from a deposit of aluminium vapour. In the final stages of making an integrated circuit, aluminium in its vapour state is allowed to form a thin layer of aluminium over the entire surface of the silicon chip.

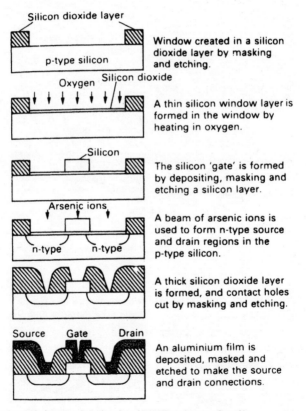

Silicon dioxide layer

p-type silicon

Window created in a silicon dioxide layer by masking and etching.

Oxygen   Silicon dioxide

A thin silicon window layer is formed in the window by heating in oxygen.

Silicon

The silicon 'gate' is formed by depositing, masking and etching a silicon layer.

Arsenic ions

n-type   n-type

A beam of arsenic ions is used to form n-type source and drain regions in the p-type silicon.

A thick silicon dioxide layer is formed, and contact holes cut by masking and etching.

Source   Gate   Drain

An aluminium film is deposited, masked and etched to make the source and drain connections.

Figure 12.9 Steps in the formation of a single n-channel MOSFET transistor on a silicon chip.

This thin layer is cut into a pattern of conducting paths using the techniques of masking and etching. When all the integrated circuits have been formed in this way, the wafer is cut up into individual chips, checked and packaged in a form that can be used by the circuit designer (Figure 12.10).

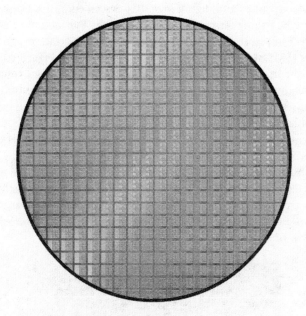

Figure 12.10 A wafer ready for cutting into individual integrated circuits.

Insight

As you can imagine, making integrated circuits that pack millions of components onto a fingernail-sized chip of silicon is rather more complex than it sounds. Imagine the problem that even a speck of dirt could cause when you're working at a microscopic scale. That's why integrated circuits are made in spotless laboratory environments called clean rooms, where the air is painstakingly filtered and workers have to wear protective clothing and pass in and out through airlocks.

12.9 Packaging silicon chips

Many silicon chips are packaged in the familiar dual-in-line (DIL) package intended for insertion into holes in a circuit board and then soldered in place if it is to become a permanent circuit. For example, op amps and digital logic ICs described in this book are of this type having terminal pins separated by 0.1 inch or 2.54 mm. However, integrated circuit packages known as surface mount devices (SMDs) are replacing this older technology. Figure 12.11 shows the use of surface mounted components in a memory stick and includes not only integrated circuits but also resistors and capacitors. These components have small metal tabs or end caps to enable them to be soldered on the surface of the circuit board. The primary advantage of this construction is that components are smaller and can be placed on both sides of the board to allow a much higher circuit density. In addition, SMDs can be made with a high degree of automation that reduces labour costs. Clearly, SMDs are designed solely for soldering into circuits whereas the dual-in-line ICs are also suitable for temporary circuit assembly on breadboard as described in Chapter 16.

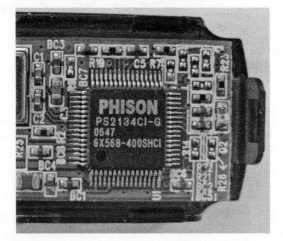

Figure 12.11 An SMD in a memory stick.

TEST YOURSELF

1 *Name the functions of four main parts of a computer.*

2 *Explain the purpose of memory in a computer system*

3 *What is the purpose of memory in a digital camera?*

4 *Describe the function and use of flash memory.*

5 *What is a memory stick and what is it used for?*

6 *Describe the functions of RAM and ROM.*

7 *What is the difference between static and dynamic RAM?*

8 *Explain the meaning of the terms photoresist and photomask in the making of a silicon chip.*

9 *What are the advantages of surface-mounted devices in the construction of electronic circuits compared with traditional dual-in-line ICs?*

10 *Explain what Moore's Law predicts in terms of the number of transistors that can be integrated on a silicon chip.*

13

....

Op amps and control systems

In this chapter you will learn:

- *the essential features of a control system*
- *that a thermostat is an example of a control system*
- *the importance of positive and negative feedback in control systems*
- *about the use of op amps as comparators in control systems*
- *the function of the inverting and non-inverting terminals of an op amp*
- *how the op amp can be used as a Schmitt trigger*
- *the basic purpose and operational details of a stepper motor*
- *the purpose and significance of a digital-to-analogue converter in a computerized control system*
- *how a digital-to-analogue converter works.*

13.1 The basic features of control systems

Control engineering is a vast field and ranges from simple thermostatic control systems for, say, the control of temperature in a tropical fish tank, to advanced position-control systems aboard spacecraft exploring the Solar System. But whatever their level of sophistication, all control systems have certain common features. The simplest form of control is open-loop control. Figure 13.1a shows its three basic elements. Building block 1 called 'desired output' is what the user of the control system wants as the 'actual output' shown by building block 3. Building block 2, the 'controller', makes the output possible after the input has been set. A typical example of an open-loop control system is a domestic light dimmer switch. The desired light level is selected by the amount a control knob is turned. However, if the light dims because of a partial power failure, there is no feedback between the output (the amount of light produced) and the input (the control setting) to enable it to make changes to the actual output once the input has been set.

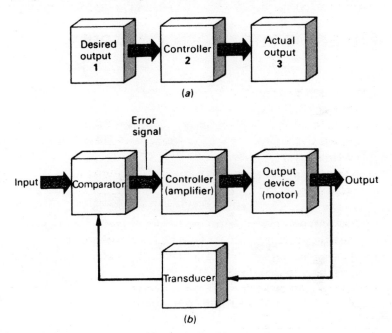

Figure 13.1 Two basic types of control system: (a) open-loop control; (b) closed-loop control.

Of course, having set the dimmer switch you could intervene to bring the light level back to what you wanted. Now you have taken action and provided feedback, the dimmer switch has become a closed-loop control system. But the system is a closed-loop control system only as long as you, the human operator, remain on duty. In automatic closed-loop control systems, a monitoring device replaces the action of the human operator, and the automatic device can't go to sleep! Figure 13.1b shows the basic elements of a closed-loop control system. The system includes a **transducer** for monitoring the prevailing state of the actual output and converting it into a form similar to the signal representing the desired output. These two signals are compared to produce a difference, or error, signal that is then used to control the system. Thus, closed-loop control systems are 'error-actuated'. Modern control systems use a variety of electrical transducers for producing the feedback signal. For example, temperature can be monitored with a thermistor, position with a variable resistor, e.g. a potentiometer, and forces with a strain gauge. (See Chapter 5 for a description of these transducers.)

Figure 13.2 An artist's impression of Spirit, one of two Mars rovers now exploring Mars under closed-loop control by NASA scientists and engineers.

Source: *http://marsrover.nasa.gov/gallery/all/spirit.html*

13.2 The design of a simple thermostat

The simple thermostat circuit shown in Figure 13.3 is an example of a closed-loop control system. It comprises three parts: a thermistor temperature sensor, Th_1, which is one component of a voltage divider; a comparator based on an integrated circuit, IC_1, called an operational amplifier; a single transistor amplifier, Tr_1, which opens and closes the relay contacts to control the power to the lamp. The detailed operation of an operational amplifier (op amp) is described below. The op amp shown in Figure 13.3 compares the voltage set on pin 3 with that on pin 2. It produces a positive voltage to switch the transistor on when the voltage on pin 2 is less than that on pin 3.

In this simple circuit, a 12 V filament lamp produces the heat. In use, the lamp could heat the air in a small box that houses tropical insects, for example. The thermistor senses the temperature in the box. Suppose VR_1 is set so that the lamp just switches on so that the output voltage from the comparator is positive. In this case, the voltage on pin 2 of the comparator is slightly lower than the voltage on pin 3. As the heat from the lamp warms the air inside the box, the resistance of the thermistor falls. When this fall makes the voltage at pin 2 rise above that on pin 3, the output voltage from the comparator falls to 0 V and the heater is switched off. Thus the temperature of the box rises or falls to a value determined by the reference voltage set on pin 3 by VR_1. The op amp provides the active sensing component that acts to 'close the loop'. A scale could be fitted to the spindle of VR_1 so that the thermostat could be used to stabilize preset temperatures once the thermostat had been calibrated.

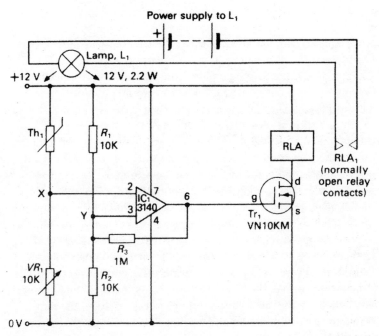

Figure 13.3 Simple thermostat circuit.

13.3 The design of a simple servosystem

A servosystem is an electromechanical device for the precise positioning of something, e.g. the positioning of a laser beam over the tracks in a compact disc player. Thus, a servosystem uses transducers to continuously monitor the position of the output, and provides corrective action to ensure that the output behaves correctly. A servosystem is an example of a closed-loop control system. Figure 13.4 shows a simple but useful example of a servosystem based on an integrated circuit op amp, IC_1. The circuit is designed to position an output shaft (attached to the d.c. motor M_1) to a desired angle by the setting of the potentiometer VR_1. This automatic setting of the output according to a preset input is provided by the mechanical link of a second potentiometer, VR_2, attached to the motor shaft. Thus, this servosystem could be used for the remote rotation of, say, a roof-top aerial to obtain optimum reception of radio signals.

A variable resistor VR_3 is used to set the sensitivity of the servosystem by altering the voltage gain of the op amp (see Chapter 14). If the gain is too high, the servosystem responds too sharply to any change in VR_1, the 'set position' potentiometer. This usually means that the output shaft will overshoot the required setting and oscillate about a mean position. These oscillations may be 'damped', dieing down after a few oscillations, or continue – a condition known as 'hunting'. If the voltage gain is set too low, the output will follow the setting of the input potentiometer very sluggishly. For optimum response, the setting of VR_3 must be such that the output reaches the desired position quickly but without overshoot. The transistors shown are suitable for controlling currents up to about 3A using heat sinks bolted to them. Since the op amp is operated from a dual power supply, its output voltage can be positive or negative with respect to 0 V. When it is positive, the npn transistor Tr_1 switches on and current flows through the motor one way; when it is negative, the pnp transistor Tr_2, switches on and current flows through the motor the opposite way.

Depending on the difference of voltage between pins 2 and 3 of the op amp, the output voltage is positive or negative. The difference is the error signal that is reduced to zero by the feedback provided by the mechanical link between the output and the input. This action is called negative feedback and is explained more fully in Chapter 14. When the error signal equals 0 V, both transistors are switched off and the motor is stationary.

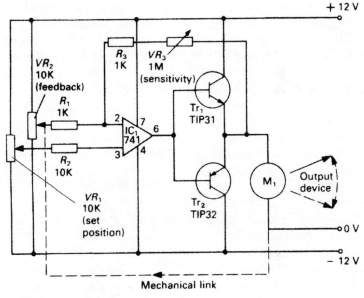

Figure 13.4 Simple servosystem circuit.

13.4 The basic properties of op amps

The principal electronic component in the thermostat and servosystem described in the preceding two sections is an integrated circuit operational amplifier or op amp. These two designs use an op amp to compare a reference voltage set on one of its two inputs with a varying voltage on its other input. The output voltage of the op amp changes abruptly when the varying voltage is slightly higher or lower than the reference voltage. This abrupt change in output voltage is used to switch a relay on or off to form the basis of a simple temperature control system or a position control system. The op amp is such an important integrated circuit in control systems and instrumentation systems that an explanation is needed about what it does, and what is particularly useful about its characteristics.

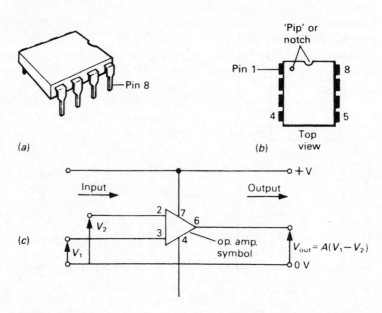

Figure 13.5 The dual-in-line (d.i.l.) op amp package: (a) its general appearance; (b) its pin identification; (c) its circuit symbol.

The appearance of one type of op amp is shown in Figure 13.5a. This is an 8-pin dual-in-line (d.i.l.) package. Figure 13.5b shows how the pins are numbered looking at the package from the top. Figure 13.5c shows the general connections to the op amp. Note its circuit symbol is a triangle, an arrowhead that shows the direction in which signal processing takes place. The op amp has two inputs (pins 2 and 3), one output (pin 6) and two power supply connections (pins 4 and 7). An op amp can be operated with one or two power supplies as shown in Figure 13.6. If it uses a single power supply, pin 4 is connected to 0 V and pin 7 to +V. Input and output signals are measured with respect to 0 V. With two power supplies, pin 4 is connected to the −V terminal of one supply, pin 7 is connected to the +V of the other supply, and the common connection of the two supplies provides the 0 V supply line. Now input and output voltages are measured with respect to this common connection.

However, why does the op amp have two inputs? Figure 13.5c shows that one input (pin 2) has an input voltage of V_2 on it, and the second input (pin 3) has an input voltage of V_1 on it. What does the op amp do with these two input voltages? It amplifies the difference between them so that the output signal V_{out} is given by the following 'op amp equation':

$$V_{out} = A(V_1 - V_2) = A \times V_{in}$$

where $(V_1 - V_2)$ is the input voltage, V_{in}, and A is known as the voltage gain of the op amp. A is the number of times the output voltage is greater than the input voltage, and it is very high. Most modern op amps have voltage gains in excess of 100,000. Note that an op amp is not sensitive to the actual values of the voltages on its two inputs – it only 'sees' the difference between them (subject, of course, to limiting maximum voltages!). The high voltage gain has to be brought under control before it is of any use in the instrumentation systems described in Chapter 14. But control systems are designed to make use of this high gain.

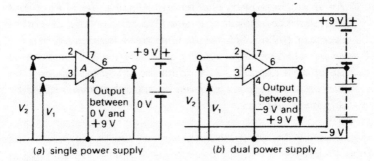

Figure 13.6 Power supply connections to the 8-pin d.i.l. version of an op amp.

One of the op amp's two inputs (pin 2) is called the inverting input and the other (pin 3) the non-inverting input. What do these names mean? Figure 13.7 shows an op amp wired up with two power supplies, but with one of the inputs connected to 0 V. The 'op amp equation' shown above can be used to find out what the output voltage is in these two cases. If a positive input voltage, V_1, is applied to the non-inverting input (pin 3) and the inverting input is connected to 0 V, the output voltage, V_{out}, is positive and equal to $A \times (V_1 - 0) = A \times V_1$. However, if a positive input voltage, V_2, is applied to the inverting input (pin 2), the output voltage is given by $A \times (0 - V_2)$. That is, $-AV_2$. Thus, any difference between the two input voltages that makes the non-inverting input voltage more positive than the inverting input provides an amplified positive (above 0 V, and therefore 'non-inverted') output voltage. Moreover, any difference between the input voltages that makes the inverting input voltage more positive than the non-inverting input voltage provides an amplified negative (below 0 V and therefore 'inverted') output voltage. It is common to indicate the inverting input of an op amp by a '−' sign to indicate its ability to invert the sign of an input signal, and the non-inverting input with a '+' sign to indicate its ability not to invert the sign of an input signal. You should not confuse these signs with the polarities of the power supplies.

Thus, we have two basic characteristics of an op amp:

1 *An op amp has a very high voltage gain, e.g. the 3140 op amp used in the thermostat of Figure 13.2 has a voltage gain in excess of 100,000. Hence op amps are sometimes referred to as 'packages of gain'.*
2 *An op amp responds to the difference of voltage between its two input terminals. Hence an op amp is sometimes called a difference amplifier.*

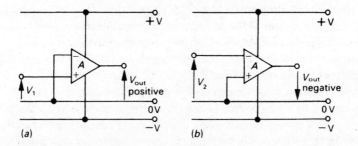

Figure 13.7 The action of (a) the non-inverting input and (b) the inverting input of an op amp.

Insight

Operational amplifiers are one of the most widely used integrated circuits today. They are used in numerous consumer, industrial and scientific devices. Their costs range from a few pence for well-known and popular op amps to over £100 for specialist devices such as instrumentation amplifiers. Op amps usually have two inputs and a single output and may be packaged singly in an 8-pin or 14-pin dual-in-line package. As explained in this chapter, high input resistance and high voltage gain are important characteristics. They can be operated with either negative or positive feedback. Negative feedback enables the designer to tailor the intrinsically high gain to a defined lower gain determined solely by external resistor values. Positive feedback, on the other hand, provides improved switching function.

13.5 Using op amps as comparators

A comparator is a circuit building block that compares the strength of two signals and provides an output signal when one signal is bigger than the other. Clearly, the op amp is one type of comparator since its two inputs are able to compare the magnitude of two voltages as shown in Figure 13.8. It simply compares a voltage, V_{in}, here shown applied to the inverting input of the op amp, with a reference voltage, V_{ref}, applied to the non-inverting input. Thus any slight difference in voltage, $e = V_{ref} - V_{in}$, causes the output voltage to saturate. Since the op amp is operated from a dual power supply, the two possible saturation voltages are $+V_{sat}$ or $-V_{sat}$. If a single power supply were used, the saturation voltages would be $+V_{sat}$ and 0 V. To 'saturate' means to go to the maximum value. In this case, the maximum output voltage is limited to a volt or two below the supply voltage, e.g. 10 V for a 12 V power supply. The output voltage cannot take on any intermediate value between the upper and lower saturation voltages since the gain of the op amp is so high. Thus if its gain is 100,000, the difference e between the two input voltages that makes the output voltage saturate at 10 V is (10 V/100000) or 0.0001 V. If the difference exceeds this value, the output voltage remains saturated at 10 V.

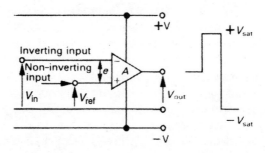

Figure 13.8 An op amp used as a comparator.

In the design of the thermostat in Figure 13.3, the reference voltage is applied to pin 3 by two equal-value resistors R_1 and R_2. The voltage divider action of these two resistors sets a reference voltage of about 6 V on the non-inverting input, pin 3. This voltage is compared with the changing voltage on the inverting input, pin 2, determined by the resistance of the thermistor. Now suppose the variable resistor VR_1 is set so that the voltage on pin 2 is lower than that on pin 3, say 5.8 V compared with 6 V on pin 3. This makes the output voltage at pin 6 rise to the upper saturation voltage, about 10 V, and transistor Tr_1 switches on. Thus the relay is energized and power is supplied to the heater. The thermistor senses this rise in temperature and its resistance falls as explained in Chapter 5. The voltage on pin 2 therefore rises. As soon as this voltage exceeds 6 V, the output voltage falls to the lower saturation voltage, 0 V in this case. The transistor switches off and power is no longer supplied to the heater. As the temperature of the thermistor rises and falls, the output voltage of the comparator falls to 0 V and rises to about 10 V respectively.

13.6 The role of the feedback resistor

There is one additional component in the design of the thermostat that 'sharpens up' its performance. This component is resistor R_3 connected between pin 6, the output, and pin 3, the non-inverting input of the op amp What does it do? Once the output voltage begins to rise or fall, this change is made even more rapid by the effect of R_3. Thus the circuit acts rather like a wall light switch, i.e. it 'snaps off' when taken past an upper position, and 'snaps on' when taken past a lower position. The electrical effect of resistor R_3 is shown in Figure 13.9. The basic comparator is shown in Figure 13.9a. Two 10 kΩ resistors, R_1 and R_2, set a reference voltage on the non-inverting input (point Y) of 6 V. The voltage on the inverting input (point X) is determined by the setting of VR_1 and the temperature of the thermistor, Th_1. Suppose this voltage is set at 6.2 V. This is higher than that at point Y, so the output of the op amp is LOW and the transistor is switched off. Now if the temperature of the thermistor falls, its resistance rises so that the voltage at point X falls. When it falls below that at point Y, the output voltage of the op amp goes HIGH thereby switching on the transistor. Note that as the output of the op amp goes from LOW to HIGH, there is no change in the reference voltage of 6 V at point Y.

Figure 13.9b shows the effect of R_3 when the output voltage is LOW. The dotted outline shows that R_3 is now effectively connected in parallel with R_2. This has the effect of lowering the value of R_2, making the reference voltage slightly lower than 6 V. Now in order for the voltage at point X to fall below that at point Y, the temperature of the thermistor must fall slightly lower than it did in the circuit at Figure 13.9a. When it does, the output voltage of the op amp rises sharply so reaching saturation. However, note the effect of R_3 on the switching action of the circuit. As soon as the voltage at the output starts to rise, the effective value of R_2 increases towards 10 kΩ. This rise in voltage raises the voltage at point Y thereby increasing the difference in voltage between points X and Y, and making the output move more rapidly towards saturation. This action is called positive feedback so named because a proportion of the rising voltage is fed back to the non-inverting input of the amp where it emphasizes the upward trend in the output voltage.

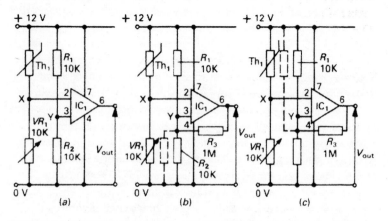

Figure 13.9 Making a Schmitt trigger from an op amp comparator: (a) the basic comparator; (b) the effect of R_3 when $V_{out} = 0$ V; (c) the effect of R_3 when $V_{out} = +V$.

Once the output voltage has risen to +10 V, R_3 is effectively connected in parallel with R_1 as shown by the dotted line in Figure 13.9c. The voltage at point Y is now slightly higher than 6 V so the temperature of the thermistor must rise slightly higher than it did in the circuit of Figure 13.9a before the voltage at point X rises above that at point Y. When this happens, the output voltage of the op amp falls. And once it starts to fall, the action of R_3 is to make the voltage at point Y fall, and that increases the difference in voltage between points X and Y thereby making the output voltage fall more rapidly towards 0 V.

The difference between the upper and lower values of the voltage at point Y is known as the hysteresis of the Schmitt trigger and is determined by the values of R_1, R_2 and R_3. It is calculated as follows.

When R_3 is in parallel with R_1 $(V_{out} = 0$ V$)$, the effective value of R_1 is 10 kΩ × 1000 kΩ/(10 kΩ + 1000 kΩ) = 9.9 kΩ. Thus the voltage at point X is 12 V × 9.9 kΩ/19.9 kΩ = 5.97 V, i.e. 0.03 V less than the value of 6 V in Figure 13.8a.

When R_3 is in parallel with R_2 (12 V, say), the effective value of R_2 is 9.9 kΩ as above. Thus the voltage at point X is 12 V × 10 kΩ/19.9 kΩ = 6.03 V, or 0.03 V more than the value of 6 V in Figure 13.8a. The hysteresis of this Schmitt trigger is therefore 0.06 V.

Insight

This 'snap-action' effect of R_3 in the performance of the circuit gives the circuit the general name of Schmitt trigger. This term is in honour of Otto H Schmitt who in 1934, while still a graduate student, first used positive feedback to improve the performance of valve-based electrical switching circuits.

If R_3 is omitted from the design of the thermostat, the relay contacts are likely to 'chatter' at the on and off switching points. This chatter is due to electrical noise that causes small changes to the voltages at the inputs of the op amp making the output voltage oscillate.

13.7 The stepper motor

This type of electric motor rotates in precise controlled steps using digital signals. Thus, a particular angle of rotation is known precisely from the number of pulses delivered to it. Stepper motors are therefore commonly used in all kinds of computer-controlled equipment, especially radio control systems, robots, computer printers and even in watches to position the hands. Their power outputs range from at least 1 kW for industrial applications to a few milliwatts when used to drive the mechanisms in model vehicles.

The principle of a permanent magnet stepper motor is shown in Figure 13.10. This has four coil windings, or phases, arranged in pairs, A, $\overline{A}$, B and B. When a d.c. voltage is applied to phase A, the resultant current sets up a magnetic field in the stator that causes the rotor to align itself with the field as shown by the arrow. If the voltage is then applied to phase B, while phase A is de-energized at the same time, the stator field will shift through 90 degrees and the rotor will move through the same angle in order to maintain its alignment with the field. Similarly, when phases A and B are subsequently energized, the rotor moves through two more 90 degree steps to complete one revolution. Of course, a stepper motor with such a big step angle as this is of little use for practical applications.

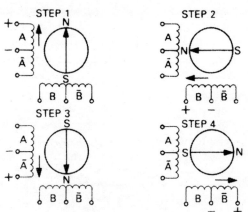

Figure 13.10 The operating principles of the stepper motor.

One common type of low-power stepper motor is shown in Figure 13.11. It is an example of a permanent magnet stepper motor since its rotor is a circular magnet with twelve pairs of magnetic poles round its circumference. Its stator comprises 24 pairs of stator poles arranged as two identical sets of 12 pairs, one set of which is offset by one quarter of the pole pitch. The rotor follows an advancing magnetic field set up when a certain sequence of pulses is fed to the stator field coils. The step angle is 7.5 degrees so it makes 48 steps per revolution.

Purpose-designed integrated circuits are available so that stepper motors can be driven by the digital signals generated by a computer control program. In its simplest form the IC would process two types of a signals from the computer: a stream of 'pulse bits', the frequency of which determines the speed at which the stepper motor rotates; and a direction bit (logic HIGH and logic LOW) that determines the direction in which the motor rotates, i.e. forward or reverse.

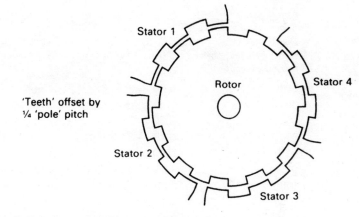

Figure 13.11 The basic constructional details of a simple stepper motor.

13.8 Digital-to-analogue converters

Being a digital control device, the stepper motor is tailor-made for use with a computer since it is operated by on/off signals that the computer is good at producing. However, it may be necessary for the computer to control the speed of analogue motors, or vary the brightness of filament lamps, or to synthesize music, which means converting digital signals into normal analogue sounds. These applications often call for the use of a special circuit building block that converts digital signals into equivalent analogue signals.

This interface device is called a digital-to-analogue converter (DAC) and Figure 13.12 shows its role as an output device in a computer control system. Note that a DAC is a type of decoder, for it decodes digital information into analogue information. The truth table of a 4-bit DAC is shown below. It gives the analogue equivalent of the sixteen values of the digital data ranging from 0 V for $(0000)_2$, to 2.25 V for $(1111)_2$. Note that a change of 1 bit in the binary information is equivalent to a change of 0.15 V in the output information. Thus if the input information is $(0100)_2$, the output information is $4 \times 0.15 = 0.6$ V and so on.

Digital input (DCBA)	Output voltage (V_{out})
0000	0.00
0001	0.15
0010	0.30
0011	0.45
0100	0.60
0101	0.75
0110	0.90
0111	1.05
1000	1.20
1001	1.35
1010	1.50
1011	0.65
1100	1.80
1101	1.95
1110	2.10
1111	2.25

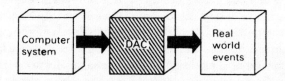

Figure 13.12 A digital-to-analogue converter is an output device in a computer system.

The principle of a simple 4-bit DAC is shown in Figure 13.13. This block diagram has two parts, a resistor network and a summing amplifier based on an op amp The resistor network takes into account that a 1 at input B is worth twice as much as a 1 at input A; and a 1 at input C is worth twice as much as a 1 at input B, and so on. Figure 13.14 shows the type of resistor network required. Each of the resistors is weighted in value in a binary sequence, i.e. R, $2R$, $4R$ and $8R$. The switches would be transistor switches in an actual DAC. They connect the resistors to the 5 V reference voltage if the input bit is a 1 and to 0 V if the input bit is a 0. Suppose the digital input is $(1101)_2$ as shown, since switches 1, 2 and 4 are connected to the 5 V reference voltage. Thus, the output voltage from the summing amplifier is given by

$$V_{out} = \frac{V_{ref} \times \left(1 + \frac{1}{2} + \frac{1}{8}\right)}{R}$$

$$= \frac{V_{ref} \times R_f \times 1.625}{R}$$

Since $V_{ref} = 5$ V, $R = 10$ kΩ and $V_{out} = 1.95$ V, $R_f = 2.4$ kΩ approximately.

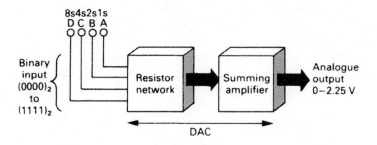

Figure 13.13 The principle of a simple 4-bit DAC.

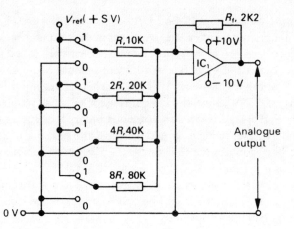

Figure 13.14 The design of a 4-bit DAC using a binary-weighted resistor network.

The problem with this simple DAC is that the resistor network requires a wide ratio of resistor values, 1 to 128, for the usual 8-bit DAC. If this simple DAC is to perform over the whole range of inputs, the resistors must be close-tolerance types whose values all vary by the same amount with any temperature change they may be subjected to. This problem is overcome by using an R–$2R$ resistance ladder as shown in Figure 13.15a. This is a technique that uses only two resistor values and that can be extended to any number of bits. Moreover, the absolute values of the resistors are unimportant since only their ratio needs to be exactly 2. The truth table below lists the output voltage, V_{out} as a function of the reference voltage, V_{ref}, for all sixteen possible values of a 4-bit binary number.

Digital input (DCBA)	Output voltage (V_{out})
0000	0
0001	$V_{ref}/16$
0010	$V_{ref}/8$
0011	$3V_{ref}/16$
0100	$V_{ref}/4$
0101	$5V_{ref}/16$
0110	$3V_{ref}/8$

Digital input (DCBA)	Output voltage (V_{out})
0111	$7V_{ref}/16$
1000	$V_{ref}/2$
1001	$9V_{ref}/16$
1010	$5V_{ref}/8$
1011	$V_{ref}/16$
1100	$3V_{ref}/4$
1101	$13V_{ref}/16$
1110	$7V_{ref}/8$
1111	$15V_{ref}/16$

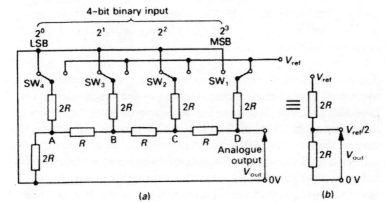

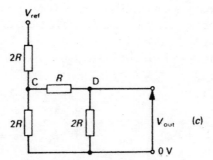

Figure 13.15 A 4-bit DAC design based on an R–2R ladder: (a) the switch positions with an input of $(1000)_2$; (b) the equivalent resistor network with an input of $(1000)_2$; (c) the equivalent resistor network with an input of $(0100)_2$.

To see how these output voltages arise, suppose $(1000)_2$ is input by setting switches SW_1 to logic 1, and SW_2 to SW_4 to logic 0. With this input signal, the entire network to the left of the node D can be replaced by a resistor of value $2R$ so that the equivalent circuit reduces to that of Figure 13.15b. The output voltage is therefore $V_{ref}/2$ as required since $(1000)_2$ represents half the full-scale input voltage.

Now input the word $(0100)_2$ by setting switches SW_1, SW_3 and SW_4 at logic 0, and SW_2 at logic 1. As shown in Figure 13.15c, a resistor of value $2R$ can replace the network to the left of node C. The voltage at node C is given by $6V_{ref}/16$, and at node D by $V_{ref}/4$, which is the output voltage. This is the voltage that is required since $(0100)_2$ represents a quarter of the full-scale input voltage. Similarly, it is possible to show that a binary input of $(0010)_2$ gives an output voltage of $V_{ref}/8$, and so on.

Figure 13.15 shows the relationship (known as a transfer function) between the digital input voltage and the analogue output voltage for this 4-bit DAC. Note there are sixteen discrete values of the output voltage for a 4-bit DAC: an 8-bit DAC has 256 discrete values of output voltage. Thus, DACs produce a stepped analogue output voltage. The more bits used, the less coarse the steps for a given output voltage and the smoother the control of the speed of the d.c. motor or of the brightness of a lamp. It is usual to use an integrated circuit package to perform the conversion from digital to analogue voltages.

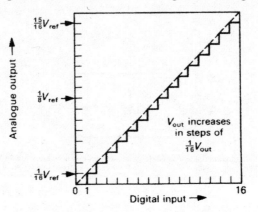

Figure 13.16 The transfer function of a 4-bit DAC.

13.9 Peripheral interface controllers (PIC)

As explained in Chapter 1, the development of the microprocessor in 1971 has led to the personal computer becoming a defining multifaceted tool of society in the 21st century. It has enabled us to access and creatively manipulate information in a wide variety of ways using standard hardware systems that can be programmed to carry out different functions. A parallel, but less obvious, revolution has taken place within electronic control system design with the programmable microcontroller (1972).

A microcontroller is a complete computer packaged in a single IC chip. With onboard memory measured in bytes and lower clock speeds than PCs, their use for data processing is limited but they can be programmed to carry out sophisticated control functions e.g. logic, timing, counting, pulsing, feedback, comparison and sequencing. This can reduce design and manufacture costs and allow circuit functions to be upgraded.

Education saw the potential for introducing students to creative electronic design and this has led to the development of a variety of easy to use programming languages. These offer an easy starting point for hobbyists with good online help. Most are based on Microchip's PIC (peripheral interface controller) range of microcontrollers and use a simple USB/stereo jack cable to download programs to the PIC.

Common programming language approaches include: Flowchart, Systems and Text. Flowchart based languages offer a good starting point. They give a visual display of the logic flow and live simulation of the sequencing which can aid fault finding. Languages such as Genie also support multi-tasking and live monitoring of PIC inputs and outputs. (Software: Genie, Picaxe, Logicator, Yenka, Flowcode)

System based languages use input, process and output system function units, such as sensors and logic gates, to be joined to form working systems. These are generally good for the design of multitasking systems but can be more tricky for sequential tasks. (eChip)

Text based languages lack a visual display of the logic flow and live simulation is usually limited, but they offer a more compact screen display and are often preferred by more experienced users. Common programming language approaches include: Flowchart, Systems and Text.

Software such as Genie and Picaxe offer free downloads. You pay for hardware or more sophisticated software options. See Taking it further for internet sources of software and hardware.

TEST YOURSELF

1 Distinguish between open-loop and closed-loop control systems.

2 By means of a block diagram explain how a thermostat works.

3 By means of a sketch explain how a servosystem works.

4 Describe the performance of an ideal op amp and state how real devices differ from this ideal.

5 Draw the symbol for an op amp and explain the meaning of its 'inverting' and 'non-inverting' inputs.

6 What is meant by 'the open-loop voltage gain' of an op amp?

7 The comparator circuit shown in Figure 13.17 is based on an op amp IC_1. The circuit uses a thermistor, Th_1, which has a resistance varying with temperature as shown in the graph. The circuit operates as a temperature-sensitive alarm and switches on the light-emitting diode, LED_1, when the temperature of the thermistor has risen to a predetermined value.
  (a) What is the voltage at point X in the circuit?
  (b) What does the graph indicate for the resistance of the thermistor at 20°C?
  (c) At what temperature does LED_1 switch on?
  (d) What would you select for the value of R_3 if LED_1 is to switch on when the temperature of the thermistor reaches 100°C?

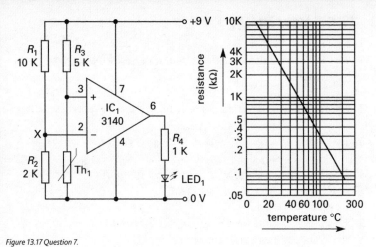

Figure 13.17 Question 7.

8 How would you modify the circuit shown in Figure 13.17 so that it works as a Schmitt trigger?

9 The circuit shown in Figure 13.18 is the design for a moisture detector to indicate when plants require watering. The three light-emitting diodes, LED_1 to LED_3, indicate 'dry', 'moist' and 'wet', respectively, depending on the resistance between the probes, P_1 and P_2. The resistance decreases as the soil in the pot becomes wetter. The three comparators are based on op amps. Each comparator compares the voltage on its inverting input (pin 2) with that on its non-inverting input (pin 3). Note that the power supply connections to the comparators are not shown for clarity.

   (a) What are the voltages at points X, Y and Z?

   (b) When the probes are pushed into dry soil, LED_1 lights. What must be the upper limit of the resistance between the probes?

   (d) Explain how the circuit makes LED_2 and LED_3 light for moist and wet soil.

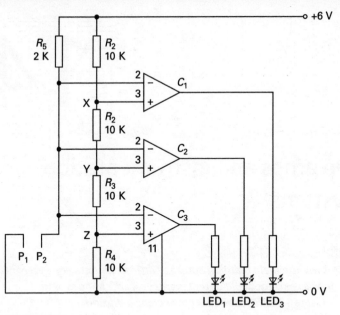

Figure 13.18 Question 9.

10 *Describe the use of the stepper motor in control applications.*

14

Op amps and instrumentation systems

In this chapter you will learn:

- **about the main building blocks of an instrumentation system**
- **how a thermocouple detects temperature changes, and a Geiger–Muller tube detects nuclear radiation**
- **how operational amplifiers are used as voltage amplifiers in instrumentation systems**
- **the equations for the voltage gain of op amps used as inverting and non-inverting voltage amplifiers**
- **about the need for analogue-to-digital converters in computerized instrumentation systems.**

14.1 Electronics and measurement

Our senses are good at detecting changes in quantities like temperature, frequency and light intensity, but instruments are needed to give us values on which we can all agree. An instrument is not only able to provide the actual value of a quantity, but it can take readings in inaccessible places, such as in the middle of a grain store and on the surface of Mars. Furthermore, instruments can measure quantities such as atomic radiation, radio frequencies and atmospheric pressure, to which our senses are quite oblivious. Moreover, digital signals propagate more efficiently than analogue signals, largely because digital signals, that are well defined and orderly, are easier for electronic circuits to distinguish from noise that is chaotic.

Electronic instruments are common, and many of these are essentially digital. Digital watches, clocks, thermometers and weighing machines are electronic. So, too, are many of the instruments used in weather forecasting, on car instrument panels and in the flight decks of airliners. The systems diagram shown in Figure 14.1 summarizes the general function of these instruments. (Also see Chapter 4.) An instrumentation system comprises three basic building blocks: sensor, signal processor and a digital or analogue display. The function of the two instruments described in this chapter, a thermometer for temperature measurement and a Geiger counter for measuring radioactivity, correspond to this system. Let us begin by looking at the characteristics of two sensors suitable for use in these two instruments.

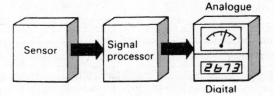

Figure 14.1 The three main building blocks of an instrumentation system.

14.2 Thermocouples and Geiger tubes

(A) THERMOCOUPLES

For accurate temperature measurement over a wide temperature range, thermocouples are generally preferred to semiconductor devices such as thermistors (Chapter 5). A thermocouple is made from two dissimilar metals or alloys, A and B, joined together to form two junctions, J_1 and J_2 as shown in Figure 14.2a.

A difference of temperature between these two junctions generates an e.m.f. Figure 14.2b shows that the variation of e.m.f. with temperature is generally linear over a wide temperature range. Commercial thermocouples are available for measuring temperature in the range –200°C to 1000°C. These thermocouples are classed as type T and type K. Type T thermocouples are based on the combination of copper and copper/nickel and operate in the range 0°C to 1100°C. Type K thermocouples are based on the combination of the alloys nickel/chromium and nickel/aluminium (like the one shown in Figure 14.2b) and are used in the range –200°C to +400°C.

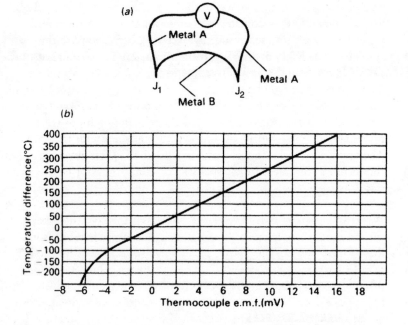

Figure 14.2 (a) The principle of a thermocouple, and (b) the e.m.f. produced by a type K thermocouple.

(B) GEIGER TUBES

Radioactive materials such as plutonium and uranium emit three main types of nuclear radiation: alpha, beta and gamma. Alpha and beta radiation are made up of particles, alpha radiation comprising the nuclei of helium atoms (two protons and two neutrons), and beta radiation comprising electrons (Chapter 2). Gamma radiation is a high-frequency electromagnetic wave (see Figure 15.2). Alpha particles have the least penetrating power, being readily stopped by paper and skin. However, alpha radiation causes damage if it comes from a radioactive substance that has somehow got into the body, for instance by breathing it in or eating contaminated food. Beta particles lose their energy within several metres in air, and they can easily be stopped by a few millimetres of aluminium. Gamma radiation has the most penetrating power but it too can be stopped by thick concrete or lead.

There are many types of sensor for detecting these different types of nuclear radiation. The one described here is the Geiger tube (strictly a Geiger–Muller, or GM, tube) and is named after Hans Geiger who lived from 1882 to 1947. What a Geiger tube counts is the passage of an alpha or beta particle, or a gamma ray, through it. A Geiger tube is very simple in principle as shown in Figure 14.3. It consists of an anode wire held at a high positive potential compared with a cathode shield that surrounds the wire. The anode and cathode are usually enclosed in a glass envelope that is filled with an inert gas such as krypton. A resistor is used in series with the anode.

When an incoming particle of radiation passes through the tube, the gas in it is briefly ionized, that is, electrons are separated from their parent atoms. This allows current to flow between the anode and the cathode which means that the cathode potential rises. One of the gases contained in the tube is a quenching agent to ensure that once the particle has passed through the tube, the potential at the anode end of the tube rises sharply to its normal value. The brief fall in potential across the tube is then amplified and fed to a monitoring circuit. Typically, the anode potential needed for the tube to work is between 400 V and 1000 V. However, precise voltage control is unnecessary in a simple Geiger counter since an individual Geiger tube has a wide operating voltage range as shown in Figure 14.4. Below about 400 V the anode potential is insufficient to allow the gas to ionize. At the Geiger threshold, the tube begins to 'count' and continues to do so over a wide range of anode potential called the plateau. As the operating potential increases, the sensitivity of the tube increases slightly. It is important not to increase the anode potential beyond the plateau since this will shorten the life of the tube considerably. A Geiger tube operated correctly remains useful for at least 500 billion counts.

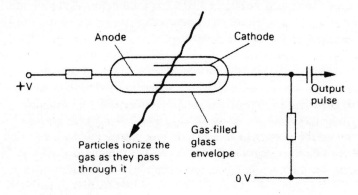

Figure 14.3 The structure of a Geiger tube.

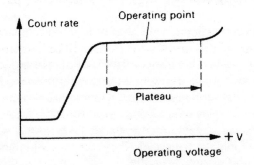

Figure 14.4 How the sensitivity of a Geiger tube varies with its operating voltage.

14.3 Designing voltage amplifiers with op amps

The basic properties of op amps and their use as comparators and Schmitt triggers in control systems are described in Chapter 13. Op amps are also used widely in the design of instruments. For example, an electronic thermometer uses an op amp to amplify the small e.m.f. that is generated by a thermocouple temperature sensor. In this application, the intrinsically high voltage gain of an op amp, that is so useful in the design of control systems, has been 'tamed' to provide a much reduced but accurately known voltage gain. How is this possible?

The technique used to control the voltage gain of an op amp is called negative feedback. (You will remember that positive feed-back is used in the design of a Schmitt trigger.) There are two basic voltage amplifier circuits making use of negative feedback: the inverting negative feedback voltage amplifier (Figure 14.5a); and the non-inverting negative feedback voltage amplifier (Figure 14.5b). In these circuits the ratio of the output voltage V_{out} to the input voltage V_{in} is the voltage gain A of the amplifier. The equations governing this voltage gain are remarkably simple as the following table shows.

Amplifier circuit	Voltage gain, A =
Inverting negative feedback voltage amplifier	$V_{out}/V_{in} = -R_2/R_1$
non-inverting negative feedback voltage amplifier	$V_{out}/V_{in} = 1 + R_2/R_1$

Note two things about these equations:

1 *First, neither equation makes reference to the intrinsically high gain of the op amp; the gain in each case is determined only by the values of the external resistors – a surprising result, perhaps? You might be tempted to suggest removing the op amp from the circuits leaving only the two resistors!*
2 *Second, the negative sign in the equation for the inverting voltage amplifier means that the input voltage is inverted, i.e. a positive input voltage produces a negative output voltage, and vice versa. The non-inverting voltage amplifier does not change the sign of the input voltage.*

Thus supposing you choose resistor values of R_2 = 100 kΩ and R_1 = 10 kΩ. In the case of the inverting amplifier the voltage gain is –100 kΩ/10 kΩ = –10 times. And in the case of the non-inverting amplifier, the voltage gain is 1 + 100 kΩ/10 kΩ = 11 times. An input voltage of +1 V produces an output voltage of –10 V in the inverting amplifier, and of +11 V in the non-inverting amplifier. Here we are assuming that the op amps are operated from a dual power supply so that the output voltage is negative, i.e. below 0 V.

Of course, any other values of resistors R_2 and R_1 could be used to obtain the voltage gain required for a particular application.

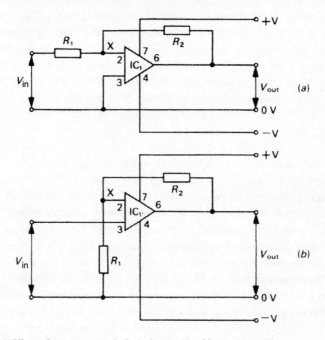

Figure 14.5 The use of op amps as negative feedback voltage amplifiers: (a) an inverting amplifier, and (b) a non-inverting amplifier.

How do the circuits shown in Figure 14.5 manage to control the intrinsically high gain of the op amp, from values of 100,000 or more (known as the open-loop voltage gain, A_{vol}), to less than 1000 (known as the closed-loop voltage gain, A_{vcl})? To explain why, we need to assume that op amps have two ideal characteristics:

1 *They have an infinitely large open-loop voltage gain, i.e. A_{vol} infinity (actual op amps have open-loop gains in excess of 100,000)*

2 *They draw no current whatsoever from the source of the signals at either of their two inputs, i.e. $I \rightarrow$ zero (actual op amps have input currents less than 10^{-6} A).*

These ideal characteristics enable us to prove the two closed-loop gain equations. First, let us start with the inverting voltage amplifier shown in Figure 14.5a in which the input voltage, V_{in}, is applied to the inverting input of the op amp via resistor R_1. Resistor R_2 is a 'feedback resistor' that 'closes the loop' between the output (pin 6) and the inverting input (pin 2). The input side of R_1 is at a voltage of V_{in}, and the output side of R_2 is at a voltage of V_{out}, both voltages measured with respect to 0 V. So what is the voltage, V_x, at the join between the two resistors, i.e. at point X, the inverting input of the op amp? Figure 14.6a shows the connections at the ends of the resistors.

Note the connection to 0 V of the non-inverting input of the op amp. Now if we have a perfect op amp (that is one that has an infinitely high open-loop gain), there is no difference between the two input voltages: if pin 3 is at 0 V, pin 2 must effectively be at 0 V. Of course, for practical amplifiers like the 741 and 3140, a very small difference of voltage (less than a microvolt) will exist between the two inputs. Thus for the perfect op amp, the voltage V_x at point X in Figure 14.6a is 0 V. This point is not actually connected to 0 V but it might just as well be. It is therefore called the virtual earth in the op amp circuit.

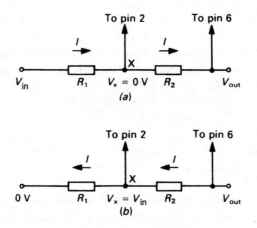

Figure 14.6 Proving the closed-loop gain equations for (a) the inverting amplifier, and (b) the non-inverting amplifier.

If we assume that the operational amplifier does not require any input current, we can concentrate our attention on working out the relationship between V_{out} and V_{in}. Let us assume that V_{in} is positive so that a current I flows through R_1 towards X. This same current flows through R_2 which is in series with it, since no current flows into pin 2. Thus, the following equations follow from the relationship between V, I and R:

> *For resistor R_1:* $I = (V_{in} - 0)/R_1$
>
> *For resistor R_2:* $I = (0 - V_{out})/R_2$

Note that the voltage difference across R_2 is $0 - V_{out}$ since current flows from pin 2 to pin 6. These equations give $V_{in}/R_1 = -V_{out}/R_2$ that can be rearranged to give

> $$(V_{out}/V_{in}) = -(R_2/R_1) = A_{vcl}$$

This is the equation that was written down above.

Now let us look at the non-inverting amplifier shown in Figure 14.5b. In this case, the input voltage, V_{in}, is applied direct to the non-inverting input, pin 3, and R_1 is connected from pin 2 to 0 V. Note that R_1 and R_2 are connected in series as shown in Figure 14.6b. The voltage at one end of R_2 is V_{out}, and at one end of R_1 is 0 V. Again, before we can use the relationship between V, I and R to prove the gain equation, we have to know the voltage, V_x, at point X between the two resistors. Since the op amp is 'perfect', there is no difference of voltage between the two inputs. So as V_{in}, changes, V_x must follow these changes; $V_x = V_{in}$. In this case, note that neither pin 2 nor pin 3 is at 0 V, virtual earth. Let us assume that a current I flows through R_1. This same current (remember there is no current into pin 2) then flows through R_2 which is in series with it. Thus, the following equations can be written down.

> *For resistor R_1:* $I = (V_{in} - 0)/R_1$
>
> *For resistor R_2:* $I = (V_{out} - V_{in})/R_2$

These equations give $V_{in}/R_1 = (V_{out} - V_{in})/R_2$. Dividing through by V_{in} and rearranging gives

$$(V_{out}/V_{in}) = 1 + (R_2/R_1) = A_{vcl}$$

This is the equation written down above.

Thus, the closed-loop voltage gain of negative feedback amplifiers is independent of variations in the characteristics of the resistors, transistors and other components making up the integrated circuit inside the op amp. It does not matter whether we use an op amp having a gain of 100,000, or 1 million, or 250,000, the closed-loop gain is determined solely by the values of resistors connected externally to the op amp. The feeding back of part of the output voltage to the inverting input is known as negative feedback and its effect is to stabilize the voltage gain of the op amp to a value determined by the values of the external resistors. Thus if the output voltage tends to rise, a small fraction of this rise is applied to pin 2. Since this is the inverting input of the op amp, the op amp acts to lower the output voltage. Likewise, the op amp counterbalances a decreasing output voltage by trying to raise the output voltage. The only stable value of the output voltage is that determined by the values of the two external resistors.

There are other advantages of using negative feedback in amplifiers like the op amp. One advantage is that it increases the bandwidth of the amplifier. As explained in Chapter 8, the bandwidth of an amplifier is the range of frequencies within which its power gain is not less than 0.7 of its maximum value. The curves in Figure 14.7 are for the 741 op amp, and show the power gain in decibels. The open-loop power gain of the 741 is about 100 dB. Now 100 dB is equal to $20\log_{10}(A_{vol})$, so $A_{vol} = 100,000$. When there is no application of feedback so the op amp is operated 'open loop' and the gain is 100 dB, the upper curve in Figure 14.7 shows that the bandwidth extends from 0 Hz, d.c., to just a few Hz. Thus, the op amp is quite useless as an audio amplifier if we wanted to use all the voltage gain it is capable of giving. But if we design a negative feedback voltage amplifier that has a closed-loop gain of, say, 40 dB (i.e. a voltage gain of 100), the bandwidth available is now about 10 kHz and the op amp is capable of giving quite a respectable performance as an audio amplifier.

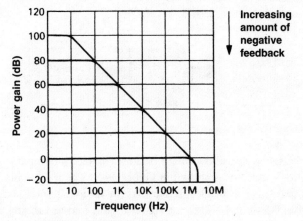

Figure 14.7 Gain versus bandwidth curves for the 741 op amp.

14.4 Analogue-to-digital converters

In Chapter 13, it was explained that a digital-to-analogue converter (DAC) is a type of decoder used at the output of digital systems, such as computers, to enable them to 'talk' to the real world where information is typically analogue. As shown in Figure 14.8, an analogue-to-digital converter (ADC) is used at the input of digital systems; it is a type of encoder that converts continuously variable analogue information, without altering its essential content, into digital information thereby enabling the digital system to 'listen in' to the real world. Thus, before a computer can 'understand' analogue quantities such as temperature, pressure and wind speed, an ADC is required. ADCs are used in all types of digital instrumentation, such as digital thermometers, multimeters and 'smart' watches that display temperature and barometric pressure. The signals from analogue sensors such as a thermocouple vary among a theoretically infinite number of values. Examples are the waveforms representing human speech passing from one person to another. In a typical telephone modem, an ADC converts the incoming audio into signals that the computer can understand. In contrast, outputs from ADCs have defined levels or states. The number of states is almost always a power of two, that is, 2, 4, 8, 16, etc.

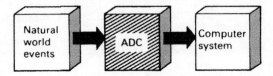

Figure 14.8 An ADC is an encoding device.

Most ADCs comprise two main building blocks, a comparator and digital logic circuits as shown in Figure 14.9. The truth table below shows that this simple design converts an analogue voltage ranging, say, from 0 V to 2.4 V into a four-bit digital output ranging from $(0000)_2$ to $(1111)_2$. Thus, an input voltage of 1.8 V is converted into a digital output of $(1100)_2$, and so on.

Analogue input	Digital output
0.00	0000
0.15	0001
0.30	0010
0.45	0011
0.60	0100
0.75	0101
0.90	0110
1.05	0111
1.20	1000
1.35	1001
1.50	1010
1.65	1011
1.80	1100
1.95	1101
2.10	1110
2.25	1111

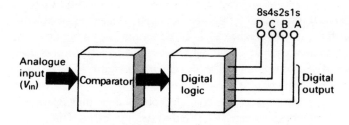

Figure 14.9 The principle of a four-bit ADC.

The principle of one type of ADC is shown in Figure 14.10a. It is called a single-ramp-and-counter ADC. The waveforms in Figure 14.10b show that the ramp generator produces an output voltage rising uniformly and then falling sharply before repeating. This ramp signal is applied to the non-inverting input of the comparator where it is compared with the steady analogue voltage on the inverting input of the comparator. While the ramp voltage is less than the input voltage, a counter counts pulses generated by the clock generator. As soon as the voltage of the ramp signal equals the input voltage, the counter stops counting and the number of pulses it has accumulated is proportional to the input voltage.

The operating sequence is shown by the waveforms. First, the counter is set to zero using a reset pulse. Then at time t_1 a 'start' pulse makes the ramp voltage rise and at the same time sets the Q output of the bistable HIGH, thereby opening the AND gate and allowing pulses from the generator to reach the counter. The counter accumulates clock pulses until the ramp voltage equals the input voltage, V_{in}. At a time t_2, the output of the comparator rises HIGH and resets the Q output of the bistable LOW. This signal closes the AND gate so that counts are prevented from reaching the counter. The digital output from the counter is then proportional to the analogue input voltage. This digital output could be fed to a computer for processing, or converted into a binary-coded decimal (BCD) format to operate a digital display as in a digital voltmeter.

The single-ramp-and-counter suffers from two drawbacks: it is slow to convert analogue voltages to digital signals, and it needs a highly stable clock generator. For faster and more stable operation, it is usual to use a dual-ramp-and-counter ADC, the principle of which is shown in Figure 14.11a. It does not need a highly stable clock generator, and a building block called an integrator is used instead of a comparator. The waveforms in Figure 14.11b show its operating sequence. At the start, switches SW_1 and SW_2 are open and the counter is reset to zero. The control logic then operates SW_1 so that the input voltage, V_{in}, is fed to the integrator. This generates a negative-going ramp that has a slope of $-V_{in}/RC$. Another building block called a zero-crossing detector sends a HIGH to the AND gate when it has detected that the ramp signal has passed through zero. The AND gate now allows clock pulses to pass to the counter.

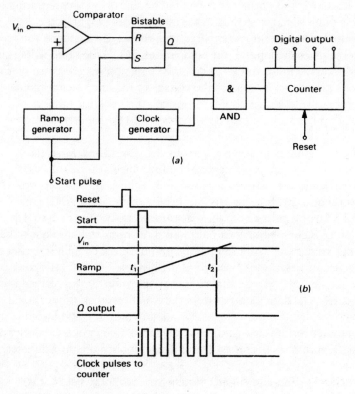

Figure 14.10 (a) The block diagram and (b) waveforms of a single-ramp-and-counter ADC.

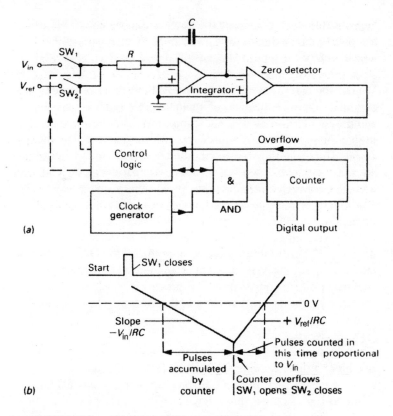

Figure 14.11 (a) The block diagram (b) waveforms of a dual-ramp-and-counter ADC.

The counter accumulates clock pulses until it reaches a maximum count of 2^n where n is the number of bits being converted by the ADC, i.e. a count of 256 for an eight-bit ADC. As soon as the counter overflows and returns to zero, it sends a signal to switch off SW_1 and switch on SW_2 so that the reference voltage, V_{ref}, is now applied to the integrator. The output of the integrator now generates a positive-going ramp that starts from the previous negative value (that was proportional to V_{in}) with a slope of $+V_{ref}/RC$. The counter begins to count until the integrator output again crosses zero. At this point, the zero-crossing detector switches LOW and closes the AND gate. The time taken for the positive-going ramp to reach zero from the previous negative value is proportional to V_{in}. Hence the number of counts accumulated in this time is proportional to V_{in}. The cycle then repeats, and a fresh conversion of the input voltage is produced.

One of the main requirements of an ADC is that it should produce a digital equivalent of an analogue voltage as quickly as possible. ADCs based on the ramp-and-counter process are inherently slow since the counter takes time to accumulate the count. The successive-approximation counter shown in Figure 14.12a is faster. It comprises a voltage comparator, a digital-to-analogue converter (DAC), a logic programmer, and a register. At the start of the conversion, the most significant bit (MSB) is applied to the DAC. The output of the DAC is compared with the analogue input voltage, V_{in}. The MSB is left in or taken out depending on the output of the comparator. If the DAC output is larger than V_{in}, the MSB is removed and placed in the next most significant bit for comparison. The process is repeated down to the least significant bit (LSB) and at this time the required number is in the counter.

Thus the successive-approximation method is the process of trying one bit at a time, beginning with the MSB as shown on the flow chart of Figure 14.12b for a four-bit conversion of an analogue input voltage. Suppose this voltage is 7 V $(0111)_2$. First, the MSB is set to 1 (block 1) and the logic circuit feeds the binary number $(1000)_2$ to the DAC. The DAC sends the analogue equivalent of this binary number to the comparator that answers the question 'Is $(1000)_2$ too high or too low?' (block 2). The answer 'Too high' is sent to the logic circuit (block 3) that cancels the MSB and sets the next MSB to 1 so that the number $(0100)_2$ is sent to the DAC and its analogue equivalent to the comparator. The answer to the question 'Is $(0100)_2$ too high or too low?' (block 4) is 'Too low'. And this is sent to the logic circuit that records the second most significant bit as a 1, and then sets the third most significant bit to a 1 (block 5). The comparison of this number with V_{in} yields the answer 'Too low', and the third bit is set to a 1 so that $(0110)_2$ is now stored. Finally a 'guess' of a 1 for the last bit sets the last bit to a 1 and finally yields the number $(0111)_2$.

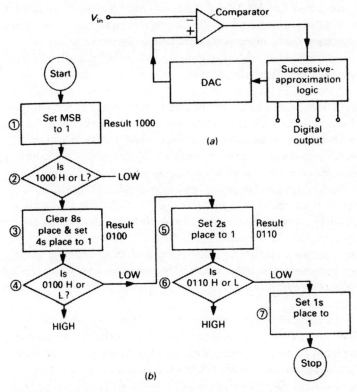

Figure 14.12 (a) The block diagram and (b) sample flowchart of a successive approximation ADC.

TEST YOURSELF

1 Name three instruments with which you are familiar and identify what aspects of their operation are 'digital' or 'analogue', or both.

2 Describe a basic electronic system for measuring temperature.

3 Explain the operation of a Geiger–Muller tube.

4 Draw the circuit symbol for an op amp and explain the meaning of its 'inverting' and 'non-inverting' inputs.

5 Name two characteristics of a 'perfect' or 'ideal' op amp.

6 Sketch the circuits for an inverting and a non-inverting amplifier based on an op amp. State the equation for the voltage gain in each case.

7 The circuit shown in Figure 14.13 is a voltage amplifier.

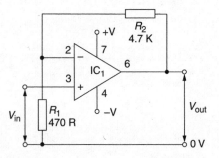

Figure 14.13 Question 7.

(a) Describe the circuit design.
(b) What is the voltage gain of this circuit?
(c) What is the value of V_{out} if V_{in} = 0.4 V?

8 *The circuit shown in Figure 14.14 is a voltage amplifier.*

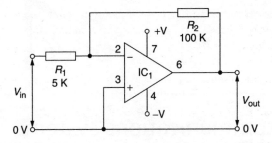

Figure 14.14 Question 8.

- **(a)** *Describe this circuit design.*
- **(b)** *What is the voltage gain of this circuit?*
- **(c)** *What is the value of V_{in} if V_{out} = 4.5 V?*

9 *Explain why the closed-loop voltage gain of an op amp used as a negative feedback voltage amplifier is independent of its open-loop voltage gain.*

10 *Explain why analogue-to-digital conversion is an important aspect of computer-based measurements.*

15

Telecommunications systems

In this chapter you will learn:

- **the general characteristics of the electromagnetic spectrum**
- **to do simple calculations concerning the frequency and wavelength of electromagnetic waves**
- **the meaning of the terms amplitude modulation, frequency modulation and bandwidth**
- **the purpose of a tuned circuit in an AM radio receiver**
- **the advantages of fibre optics in telecommunications systems**
- **the advantages of digital radio and TV compared with older analogue systems.**

15.1 Introduction

Telecommunications began when the electric telegraph was invented at the start of the 19th century. It has become the world's fastest-growing industry and includes radio, television, optical communications, GPS and mobile phones. All telecommunications systems have one thing in common: the messages they send are converted into signals that can be transmitted through wires, interplanetary space and even glass fibres.

Figure 15.1 shows the basic building blocks of a telecommunications system. It comprises a communications channel along which a message is transferred between a transmitter and a receiver. In the transmitter, a transducer (e.g. a microphone) takes the message from a source and converts it into a suitable form for transmission along the communications channel. In the receiver, another transducer (e.g. an earphone) delivers a copy of the transmitted information to a destination. Other building blocks might be part of this communications system. For example, the distance possible with the system could be improved using an amplifier to boost the strength of the signal from the microphone. Similarly, a signal weakened during transmission could be boosted using an amplifier in the receiver. A more sophisticated system might employ an encoder in the transmitter and a decoder in the receiver to give signals suitable characteristics for transmission. For example, Morse code is a way of encoding a language, such as English, into long and short pulses of radio waves for effective transmission to a receiver that decodes back into the original message.

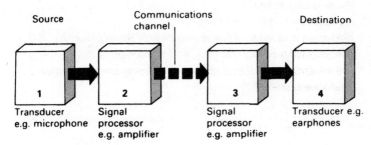

Figure 15.1 The building blocks of a communications system.

Insight

Note that the system shown in Figure 15.1 applies equally well to telecommunications systems that are not electronic. For example, when a honey bee has information about the source of a good supply of pollen, it does a special dance (using its body as a transducer) to encode this information into chemical and visual patterns. Other hive-dwellers decode these patterns into the original information about the distance and direction of the source of the pollen.

(Contd)

Figure 15.1 oversimplifies the process of gathering, transmitting and receiving information (and that is especially true of the dance of the honey bee), but it will serve as a model on which to base our understanding of radio, television and optical communications systems.

15.2 The electromagnetic spectrum

The radio, television and light waves that are used for sending messages from one place to another are part of the electromagnetic spectrum. As Figure 15.2 shows, this spectrum extends from radio waves to gamma rays. It includes X-rays, infrared and ultraviolet rays, and the visible spectrum comprising red, yellow, green, blue, indigo and violet light. Now there are two properties that all these waves have. They all:

- *travel at a speed of 300 million metres per second (3×10^8 ms^{-1}) in a vacuum (and very slightly slower in air)*
- *consist of oscillating magnetic and electric fields – hence the name electromagnetic waves.*

The variation in the strengths of the fields associated with electromagnetic waves is shown in Figure 15.3. This variation is known as a sinusoidal waveform and it has four main characteristics:

1 *an amplitude, A, that is related to the strength of the electric or magnetic field;*
2 *a wavelength, λ, that is the distance between two consecutive parts of the waveform that have the same amplitude, e.g. from peak to peak;*
3 *a frequency, f, that is the number of complete waveforms that pass a point in one second; and*
4 *a speed, c, that is the distance the wave moves in one second.*

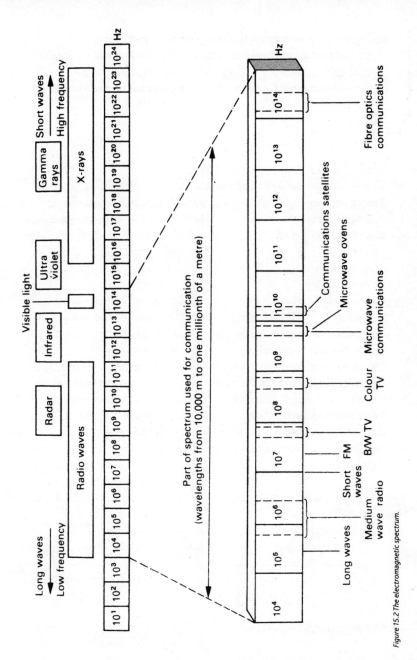

Figure 15.2 The electromagnetic spectrum.

Now there is a simple relationship between the speed of an electromagnetic wave and its frequency and wavelength. This is:

Speed of an electromagnetic wave = frequency × wavelength

Or, in symbols,

$$c = f \times \lambda$$

This equation can be used to work out the wavelengths of some of the waves shown in Figure 15.2. Thus, medium wave radio waves that have a frequency of, say, 1 MHz (one million hertz) have a wavelength given by

$$\lambda = c/f = 3 \times 10^8/10^6 = 300 \text{ m}$$

You may have noticed that this wavelength, or its equivalent frequency of 1 MHz, is marked near the centre of the tuning scale on medium waveband radios.

It is not necessary to discuss the complex nature of the electric and magnetic fields that are the very essence of electromagnetic waves, except to note that these fields oscillate at right angles to each other and to the direction of the wave as shown in Figure 15.4. It is usually only the variation of the electric field strength that is shown on waveforms of electromagnetic waves.

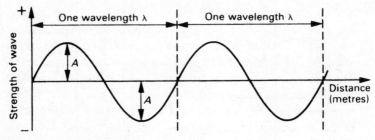

Figure 15.3 Features of a sinusoidal waveform.

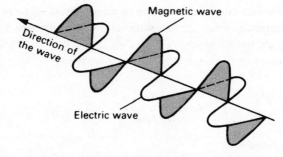

Magnetic wave

Direction of the wave

Electric wave

Figure 15.4 Electric and magnetic waves make up an electromagnetic wave.

15.3 Uses of radio waves

In the electromagnetic spectrum, radio waves extend from a frequency of about 30 kHz (30 × 10³ Hz) to more than 3 GHz (3 × 10⁹ Hz). This region is divided into the main frequency bands (shown on page 272).

Depending on the frequency of radio waves, the shape of the aerial used and the power of the transmitter, radio waves reach a receiver by one or more routes as shown in Figure 15.5. Surface or ground waves follow the curvature of the Earth. The range of these waves is limited because poor conductors such as sand absorb them, but they travel further over water since it is a better conductor. The range may be about 1500 km for long waves having a frequency less than about 300 kHz, but only a few kilometres for very high frequency waves. The sky wave travels upwards from an aerial, but if its frequency is less than a frequency of about 30 MHz, it is returned to Earth by the ionosphere. On reaching the ground, the wave is reflected back to the ionosphere, and so on until it is completely attenuated (weakened). The ionosphere consists of several layers of positively charged ions and electrons. These are produced by ionization of gases in the upper atmosphere by the Sun's ultraviolet radiation. The various layers of the ionosphere extend between about 50 km and 500 km above the Earth, and their intensity, i.e. amount of ionization, and altitude vary with the time of day, the seasons and the 11-year sunspot cycle.

Low, medium and high frequency radio waves have a range of several thousand kilometres and can travel round the world by repeatedly 'bouncing off' the ionosphere.

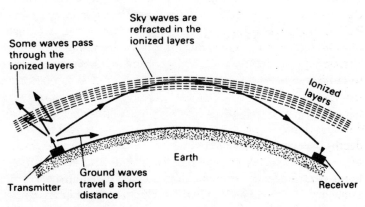

Figure 15.5 The different paths taken by radio waves.

Radio waves that have frequencies greater than about 30 MHz can penetrate the ionosphere. Indeed, communications with most satellites and interplanetary spacecraft take place at frequencies in excess of 1000 MHz (1 GHz). At frequencies of about 100 GHz radio waves begin to be absorbed again by oxygen and other gases in the atmosphere.

Radio waves are produced quite naturally by lightning flashes in the atmosphere of our own planet, but with much greater intensity in the atmospheres of planets Jupiter and Saturn. This background of radio waves from space is not just of academic interest, for it does sometimes interfere with regular radio communications on Earth. Indeed, the Sun is a particularly strong source of 'radio interference', especially when there is increased sunspot activity at the peak of the 11-year sunspot cycle.

It is in the radio 'window' between 30 MHz and 100 GHz that radio astronomers are able to study the radio emissions from space.

The Sun, the stars, dust clouds, certain types of galaxies called quasars, black holes and other exotic deep space objects produce enormous amounts of radio energy. One such radio source is the Centaurus-A radio galaxy shown in Figure 15.6. In fact, there is a whole branch of astronomy concerned with trying to understand the processes in deep space that give rise to these radio waves, as well as to gamma rays, X-rays and infrared. One exciting branch of radio astronomy is called SETI (short for Searching for Extra-Terrestrial Intelligence) and uses large radio telescopes to 'listen in' to radio energy from deep space in the hope of receiving messages from distant civilizations! The United Kingdom's biggest radio telescope is the Lovell Telescope at Jodrell Bank shown in Figure 15.7. Since the summer of 1957 it has been quietly probing the depths of space, in a search to understand the universe in which we live. Unfortunately (although that depends on your point of view!), radio astronomy has not found any extraterrestrial messages from intelligent life – or even from a time-travelling Starship Enterprise!

Figure 15.6 Centaurus-A. This spectacular galaxy is about 15 million light-years from Earth and is a powerful source of radio emission. The photo shows that it has an ellipsoidal body crossed by a curious, complicated lane of cool, dense dust that may have been left over after two galaxies collided. Centaurus-A also produces a huge amount of energy at X-ray and optical wavelengths.

Courtesy: Anglo-Australian Observatory

Figure 15.7 The Lovell Radio Telescope in Cheshire, England, still operating after over 50 years. This telescope is linked with other radio telescopes to provide high resolution maps of cosmic radio sources. The telescope is dedicated to Sir Bernard Lovell, the pioneer of radio astronomy in the UK.

Courtesy: Joddrell Bank Observatory Science Centre

Insight

If astronomers are to keep in touch with an interplanetary spacecraft as it explores the Solar System, complex electronic receiving and transmission equipment is needed on board the spacecraft. A milestone was reached in long distance interplanetary communication when the *Voyager 2* spacecraft, launched from Earth in 1977, was switched on by radio waves from Earth to send back intriguing information about the outer planets Jupiter, Saturn and Uranus, and it reached Neptune in 1989. This spacecraft is now in the remote regions of the Solar System and contact with it will continue for some time yet. However, communication with this highly successful spacecraft will be lost well before it reaches another star (or alien intelligence?) in about 350,000 years!

15.4 AM and FM modulation

It is rather difficult to produce radio waves at audio frequencies, the frequencies of speech and music, since they require a lot of electrical energy, and larger transmitter aerials are needed as the wavelength of the radio waves increases. For example, consider one such simple aerial, the dipole (or Yagi) aerial shown in Figure 15.8. It consists of two horizontal or vertical conducting rods or wires fed with radio energy from the centre. A vertical dipole like this radiates energy equally in all directions. Now if this aerial is to operate at optimum efficiency, each rod must have a length one quarter of the required wavelength ($\lambda/4$). Thus, using the equation velocity = frequency × wavelength, or $c = f \times \lambda$, the length of each rod of this dipole aerial would have to be about 7.5 km long for an audio frequency of, say, 10 kHz,

$$\lambda = c/f = 3 \times 10^8 \, \text{ms}^{-1}/10^4 \, \text{s}^{-1} = 30,000 \, \text{m}$$

Thus,

$$\lambda/4 = 7500 \, \text{m, or } 7.5 \, \text{km}$$

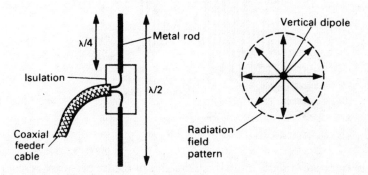

Figure 15.8 The dimensions and radiation pattern of a dipole aerial.

For frequencies above 10 kHz, it becomes increasingly easier to produce radio waves. Along with the increase in frequency, shorter aerials can be used for both transmission and reception. Thus, data transmitted from a weather satellite at a frequency of 135 MHz (wavelength 2.22 m) can be received on a dipole aerial made of two rods each 0.55 metres long.

A process called modulation is used to enable these high frequency radio waves to carry audio frequency information. There are two main processes for modulating a radio frequency carrier wave: amplitude modulation (AM) and frequency modulation (FM).

In amplitude modulation, the amplitude of the carrier wave is made to follow the variations in the amplitude of the audio frequency wave as shown in Figure 15.9. The modulation depth, m, of this AM radio wave is defined as a percentage as the ratio $(A/B) \times 100\%$. If this ratio exceeds 100%, the audio frequency message will be distorted. Too low a modulation depth produces poor quality sound at the receiver. A value of 80% is satisfactory.

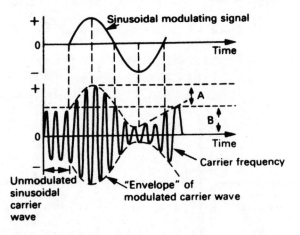

Figure 15.9 An amplitude modulated carrier wave.

In frequency modulation, the frequency of the carrier wave is made to follow the amplitude of the audio frequency waves as shown in Figure 15.10. When the amplitude of the audio frequency wave is zero, the carrier wave has a particular frequency (the frequency marked on the tuning scale of an FM receiver). An increase in the amplitude of the audio frequency wave in the positive direction produces a slight increase in the frequency of the carrier wave. An increase in amplitude in the negative direction produces a decrease in the frequency of the carrier wave. Note that the amplitude of the FM wave remains constant, so allowing the transmitter to operate at high efficiency.

The main advantage of FM over AM is that the noise level at the receiver is reduced. Any electrical noise, e.g. from lightning and electrical machinery, tends to amplitude modulate the carrier and this appears at the output of the receiver. But an FM receiver is only sensitive to the frequency variations of the carrier wave not amplitude variations. Therefore, most of the electrical noise is eliminated.

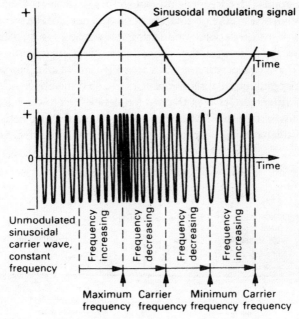

Figure 15.10 A frequency modulated carrier wave.

15.5 Bandwidth

In general terms, bandwidth is the frequency range of input signals to which an electronic system responds. Thus, you may recall (Chapter 8) that the bandwidth of an audio amplifier is the range of frequencies it amplifies as measured between the 3 dB points on the frequency response graph. Similarly, the bandwidth of a telecommunications system is the capability of that system to transmit signals of different frequencies. As you will see, the bandwidth of a telecommunications system is directly related to the information capacity of that system.

It is possible to show mathematically that when a sinusoidal carrier wave is amplitude modulated (Figure 15.9), the modulated carrier wave contains three frequencies. Figure 15.11a shows that one of these frequencies is the original carrier frequency, f_c. The second is the sum of the carrier and modulating frequencies, $f_c + f_m$. The third is the difference between the carrier and modulating frequencies, $f_c - f_m$. These two new frequencies are called sidebands. The sum of the carrier and modulating frequencies is called the upper sideband, and the difference between the two frequencies the lower sideband. The bandwidth of this AM carrier wave is the difference between the two sideband frequencies, or twice the modulating frequency:

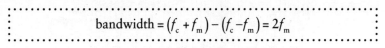

$$\text{bandwidth} = \left(f_c + f_m\right) - \left(f_c - f_m\right) = 2f_m$$

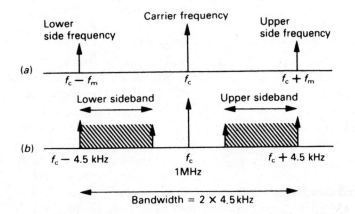

Figure 15.11 (a) The formation of side frequencies, and (b) sidebands in an AM radio signal.

When the modulating signal consists of a band of frequencies, as in speech and music, each individual frequency produces upper and lower sideband frequencies about the unmodulated carrier wave.

This results in upper and lower sideband frequencies as shown in Figure 15.11b. Thus, if the maximum frequency of the modulating signal is 4.5 kHz, the bandwidth required is 9 kHz centred on a carrier frequency of, say, 1 MHz. The higher the modulating signal bandwidth, the wider is the modulated signal bandwidth. It is obvious that the transmission system used must be capable of handling the bandwidth required of it. For example, in order to avoid overcrowding on the AM medium waveband extending from about 300 kHz to 3 MHz, each station is allocated a 9 kHz bandwidth.

Sidebands are also produced when a carrier wave is frequency modulated. However, the arrangement of the sidebands is more complex than with amplitude modulation. A number of sidebands are produced, not just two, and the way they are distributed and the number of them is complex. The number of sidebands is determined by the ratio

$$\frac{\text{maximum variation of carrier frequency}}{\text{modulating frequency}}$$

and is known as the modulation index.

The maximum variation of carrier frequency is determined by the performance required. For the BBC's FM sound broadcasts on the VHF radio band, this is 75 kHz. If the modulating frequency is 15 kHz (about the maximum transmitted) the modulation index is 75 kHz/15 kHz, which is 5, and the number of sidebands is 16 (8 on each side of the carrier). These are all spaced 15 kHz apart so the total bandwidth is 240 kHz. Moreover, the bandwidth is dependent on the magnitude of the modulating voltage whereas in AM broadcasts the bandwidth is fixed. Thus, the bandwidth required by FM, 240 kHz, is much greater than that required by AM, 30 kHz, if the modulating signal has a frequency of 15 kHz. This higher bandwidth necessitates the use of higher carrier frequencies and that is why FM sound broadcasting is on the VHF band – see Figure 15.2.

15.6 Radio transmitters and receivers

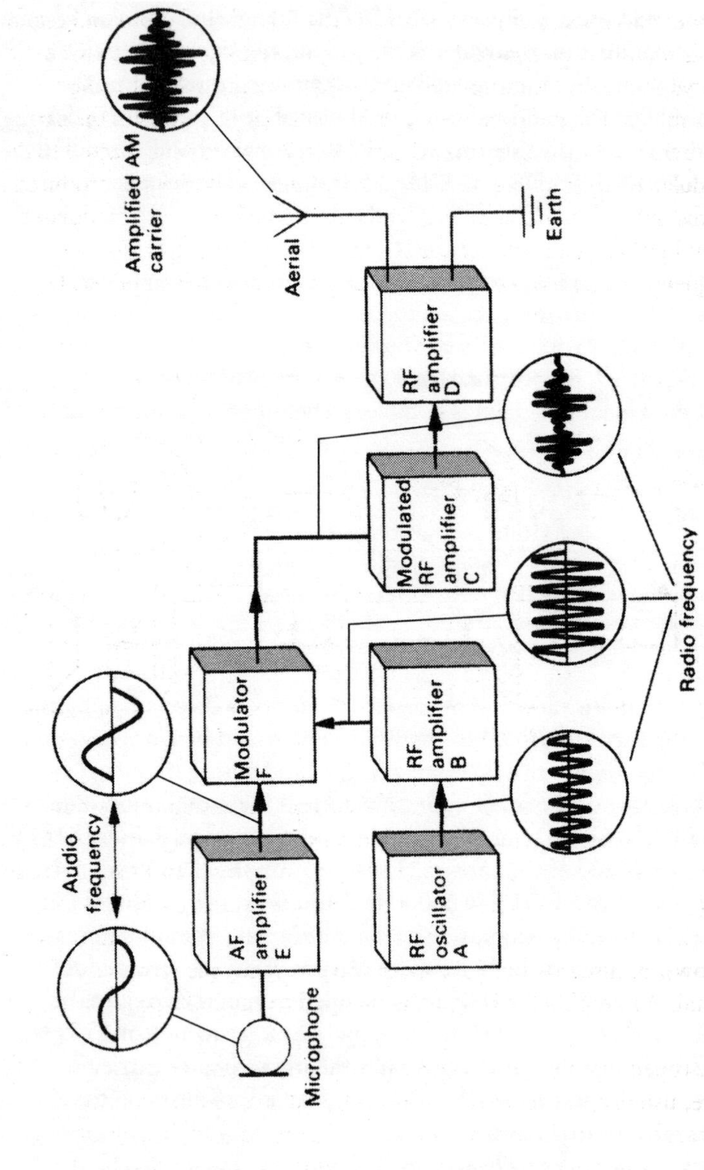

Figure 15.12 The building blocks of an AM transmitter.

The prevailing radio communications systems in use today are analogue in that they process signals that are continuous and smoothly changing. However, the growth of digital communications for mobile phones and consumer goods such as the DVD, digital radio and digital TV is rapidly challenging this supremacy (see below). Figure 15.12 shows the main building blocks needed to make an AM radio transmitter. The radio frequency (RF) oscillator, A, generates the carrier wave that is amplified by the RF amplifier, B, before being passed to the modulated RF amplifier, C. The audio frequency (AF) signal produced by the microphone is amplified by the AF amplifier, E, and modulator, F, and passed to the modulated RF amplifier, C. An AM radio frequency signal is fed to the aerial via the optional RF amplifier, D.

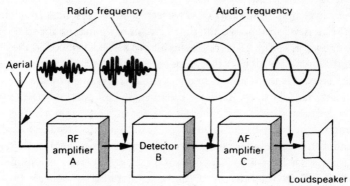

Figure 15.13 The building blocks of a tuned radio frequency receiver.

The AM radio frequency signal can be received using the system shown in Figure 15.13. An RF amplifier, A, selects a particular AM carrier wave and amplifies it. The selection is done by a tuned circuit (see below) and amplified by an RF amplifier having a bandwidth of 9 kHz to accept the sidebands. A detector (or demodulator – see below), B, discards the RF carrier wave to leave the required AF signal. An amplified AF signal is then fed to the loudspeaker.

The tuned circuit of an AM radio receiver comprises a coil of wire, usually wound on a ferrite rod, and a capacitor connected in parallel with the coil as shown in Figure 15.14a. This building block resonates at a characteristic frequency determined by the values of the inductance, L, of the coil and the capacitance, C, of the capacitor. Inductance is measured in units of henries (H) while capacitance (Chapter 6) is measured in farads (F). In an AM radio receiver, the inductance of the coil would be a few microhenries and the capacitance of the capacitor a few hundred picofarads. By using a variable capacitor, different stations can be tuned in by making the tuned circuit resonate at different frequencies. The resonant frequency, f_o, is given by the equation

$$f_O = 1/(6.28\sqrt{LC})$$

This equation indicates that increasing the values of C and/or L reduces the frequency at which the circuit resonates, i.e. increases the wavelength of the AM waves selected. At the resonant frequency, the p.d. developed across the tuned circuit has a maximum value as shown in Figure 15.14b. Negligible p.d.s are developed in the tuned circuit for all the other frequencies that the aerial is picking up. Thus, it is the resonant frequency that is amplified by the circuit that follows it. Actually, the tuned circuit does not select just one frequency but a narrow band of frequencies as shown in the graph. Obviously if the tuned circuit is to 'separate' closely spaced stations, the bandwidth must be small – but not too small since the carrier wave from a particular station requires a certain minimum bandwidth in order to carry a full range of audio frequencies. In Figure 15.14b, the bandwidth is defined as the difference between the higher and lower frequencies, f_h and f_l, where the p.d. across the tuned circuit has fallen to 0.7 of the maximum p.d.

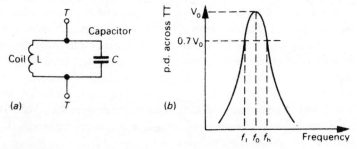

Figure 15.14 (a) A parallel tuned circuit, and (b) the p.d. across it at its resonant frequency.

The process of detection or demodulation of an AM radio wave can be achieved with the circuit shown in Figure 15.15. During each positive part of the radio frequency signal, diode D_1 passes a current that charges up capacitor C_1. As the positive peaks subside, the capacitor discharges through R_1. If the capacitor voltage remains close to the peak voltage until the next peak of the radio frequency signal arrives, only the envelope of the AM signal (that is the audio frequency modulation) is passed by the detector; i.e. the detector is a low-pass filter. This means that the time constant $R_1 \times C_1$ must be between one period of the RF signal (say, 10^{-6} s for a radio frequency of 1 MHz) and one period of the AF signal (say, 10^{-3} s). If the time constant is too large, C_1 discharges too slowly and the output voltage does not follow the AF signal, i.e. the detector is insensitive. And if the time constant is too small, the RF signal passes through the demodulator. Typical values are $C_1 = 0.01$ µF and $R_1 = 10$ kΩ giving a time constant of 10^{-4} s.

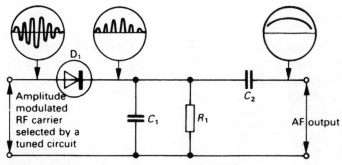

Figure 15.15 The principle of demodulation.

A simplified systems diagram for an FM transmitter is shown in Figure 15.16. The audio output from the microphone is amplified by the radio frequency amplifier, A, and fed to the RF oscillator, B. This produces a frequency modulated signal that is fed to the RF amplifier, C, and then to the aerial. Note that the RF amplifier increases the amplitude of the signal but does not alter the frequency.

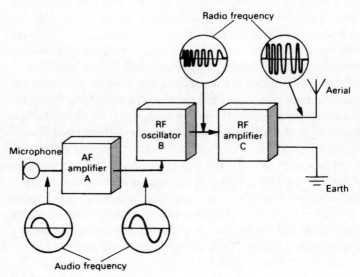

Figure 15.16 The building blocks of an FM transmitter.

These FM radio signals can be received with the system shown in Figure 15.17. This is basically the same as for the reception of AM radio signals except that the demodulator has to respond to the variations of the carrier frequency instead of variations of carrier amplitude.

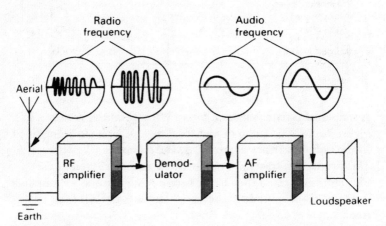

Figure 15.17 The building blocks of an FM receiver.

15.7 Digital radio and TV

Digital radio is the transmission and reception of sound that has been processed using patterns of the binary digits, 1s and 0s. Because the sound is transmitted as a digital signal, digital radio receivers can provide sound quality approaching that of CDs. Unlike analogue AM, digital radio reception eliminates the noise that often invades analogue radio transmission and reception. Moreover, unlike analogue stations, national digital radio stations are broadcast on the same frequency across the country so it is unnecessary to retune receivers in different parts of the country. Thus, if you have a digital radio in your car you do not need to retune while driving as it monitors signal strengths and uses this information to switch automatically from a fading signal to a new, more powerful one that you may be approaching. A digital radio is a 'smart' radio in that its computer takes over the tuning by switching from one transmitter to another without you noticing any break in the signal. However, apart from these advantages of digital radio whether the quality of speech and music is any better than FM reception is debateable.

Digital radio in the UK is based on the Eureka 147 Digital Audio Broadcasting (DAB) system that was developed in Europe by a consortium of broadcasters and manufacturers. This system is an accepted European Telecommunications Standard and is likely to be the standard system throughout Europe, Canada, Australia, New Zealand, Russia, China, India, South Africa, Mexico, Malaysia and Singapore. The Eureka 147 system uses digital compression that transforms the source material, i.e. music, speech and radio programmes, into digital computer code before it is stored or transmitted.

In the UK, seven blocks of frequencies in Band 3 have been assigned for use by digital radio. They lie between 174.982 MHz and 239.200 MHz, but are more generally known as 11B, 11C, 11D, 12A, 12B, 12C and 12D. Each frequency block can carry one digital radio multiplex. A multiplex is just a fancy word for a number of radio services mixed up together. Each multiplex can carry up to six or more services, and this is one of the key differences between analogue and digital radio. In analogue radio, one frequency means one service; but in digital radio, one frequency block can carry many services, thus making efficient use of the radio spectrum.

Stations are usually listed by name on the receiver and are easily tuned in at the touch of a button. Because the digital radio receiver is 'smart', it can do much more than just pick up radio signals. For example, there is no need to select frequencies because the station you want is selected from the call letters, or names displayed sequentially on its LCD screen, and the computer within the radio does the rest. This display may well present other information such as song titles, artists and album names and lyrics; traffic and weather information, including emergency warnings; text services, such as stock market quotations; or complementary information about a product being advertised on-air.

As with digital radio, the transmission of pictures digitally provides many advantages over analogue television that has been in use since the 1930s. The technical quality can be much better and more consistent because the information needed to make up the TV programme is coded into a digital data stream. The digital stream takes up much less capacity in the airwaves, so that the space needed in the past for just one analogue channel can now carry five, six or seven different programmes. This means a much greater choice of services to watch if you are a digital viewer. Moreover, digital TV can be accommodated in widescreen format so as to be compatible with the programmes that are in true widescreen – something analogue TV cannot do properly. The widescreen format TV having a 16:9 aspect ratio was introduced to provide viewers with a more natural perspective than the common 4:3 aspect ratio, and reflects Hollywood's choice in the 1950s for *Cinerama*, *Cinemascope* and *VistaVision*.

High definition TV (HDTV) combines the widescreen format with digital communications to give more definition on screen than on a standard TV screen. Each pixel is still made up of three close dots of coloured phosphors of red, green and blue, but the pixels are smaller and closer together, and square-shaped rather than rectangular, just like most colour computer monitors.

Although you can still watch TV passively, digital TV also allows you to make the viewing experience your own by accessing related information running alongside the programme you are watching. Each channel has about 19.2 megabits per second of data that can be added to each broadcast. Most of it is video and audio, but some of the signal can be other forms of data just like a very fast network connection sending pictures, sound, multimedia games, and illustrated articles, all related to the television program you are watching.

Recently, the sales of liquid crystal display televisions (LCD TV) have exceeded sales of conventional CRT televisions worldwide. One reason for the surge in interest is that LCD televisions are thinner and lighter than the CRT-based TVs of similar display size, and are also available in much larger sizes. This combination of features made LCDs more practical than conventional TVs for many roles, not least for saving space in the living room.

Insight

It seems that holographic TV is round the corner – well, in 10 years anyway. Displaying three-dimensional images, holographic TV means that the animals in a nature programme would appear in your living room as if you were filming them in real life. Holograms are interference patterns of light generated by the interaction of a uniform reference laser beam with a second beam that has been reflected from an object of interest. If a beam similar to the original reference beam is the shone through a hologram, the result is a three-dimensional image of the object scanned. Just as a moving picture is actually a series of stills shown in quick succession, a moving hologram would be a series of still holograms to fool the brain. The problem is that individual still holograms contain so much information that they require a special medium to record them and vast computing power.

15.8 Plasma and LCD TV screens

These terms represent two competing display technologies, both producing crystal clear images and both coming in shallow flat screen casing. In the case of a plasma flat screen, each pixel is like a microscopic fluorescent lamp which receives instructions from software contained on the rear electrostatic silicon board. Electric pulses excite natural gases such as neon and xenon causing the pixels to glow. This light illuminates the proper balance of red green and blue phosphors, all contained within each cell, to display the correct colour sequence from the light. If you look closely at a plasma TV you will see that each individual pixel is made up of red green and blue bars. In contrast, an LCD display comprises a matrix of thin film transistors (TFTs) that supply voltage to liquid crystal from cells that are sandwiched between two sheets of glass. When excited by an electrical charge, the crystals untwist by a precise amount in order to filter white light generated by a lamp behind the screen if it is a flat-panel TV. LCD monitors produce colour by a process of subtraction whereby particular colour wavelengths from a spectrum of white light is blocked out until you are left with the right colour.

So what are the advantages of each one? Well, plasma screens produce better picture quality in low-light conditions, while LCD screens are better for public display in a bright room environment. The latter are particularly good for use in computers since the number of pixels per square centimetre is generally higher than for plasma. On the other hand, plasma TV screens are better at responding to fast-moving images as in video playback, LCD screens tending to show a 'tail' where individual pixels are just out of step with the image on the screen. There is good reason for using LCD panels in aircraft since they are unaffected by the increases or decreases in air pressure. This is not the case for plasma panels since the display element is actually a glass substrate envelope containing rare natural gases. The effect is that, at altitude, plasma displays emit a buzzing sound due to the lower air pressure.

A number of manufacturers have developed so-called LED TVs resulting in a thinner TV. Also, since the images displayed have 100 Hz display rate rather than the usual 50 Hz, this reduces blur in fast-moving scenes. However, the LEDs are not in themselves the pixels but simply replace the ordinary fluorescent backlights with multiple LEDs that brighten, dim and turn off altogether depending on the picture on screen. This gives greater subtlety of colour. It occurs to me that this quest for higher definition TV images in more vibrant colours will eventually result in images that no longer represent reality but which are an artificial rendering of it.

15.9 The CCD camera

In the traditional analogue cameras is a different type of camera from the Vidicon tube which is used to for converting light into electrical signals since it is based on semiconductors rather than on thermionic emission. The CCD comprises a light-sensitive array of metal-oxide semiconductors on a silicon chip. It is much more sensitive than the Vidicon tube and it is also able to store images in such a way that each pixel (picture element) in the image is converted into an electrical charge, the intensity of which is related to the colour in the colour spectrum. The charge is removed from the array and processed to form an image by electrodes attached to the surface of the chip. The longer the light falls on the array, the larger the charge stored. Thus, the CCD is more sensitive than photographic film and enables light to be gathered from faint sources. For example, both amateur and professional astronomers use CCD cameras in place of the eye and photographic film to record images in their telescopes. For surveillance purposes, CCD images can be gathered in near darkness, and CCDs are used in digital cameras, camcorders, scanners and barcode readers. The word **megapixel** refers to the remarkable resolution (the amount of detail) of images taken by CCDs in digital cameras, one megapixel being an image comprising, say, 1024 × 1024 pixels.

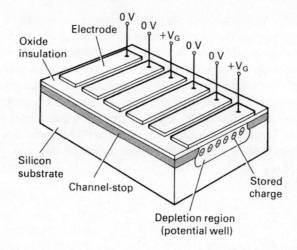

Figure 15.18 Cross-section through a CCD.

The way a CCD works is shown in Figure 15.18, which provides a cross-sectional view of the semiconductor chip making up the CCD. On the surface of the chip is a long row of tiny metal spots (electrodes) that overlay a thin oxide layer formed on the surface of a p-type substrate. A three-phase clock network alternately activates the electrodes in turn by being switched from 0 V to say +10 V. When an electrode is pulsed to a positive voltage, it is capable of attracting a negative charge to the underside of the oxide layer beneath it. It is as if the positively charged metal electrode creates a kind of 'bucket' that can hold electric charge.

Charge-coupling is the technique by which signal charge can be transferred from the bucket under one electrode to the next bucket. This is achieved by taking the voltage on the send electrode also to +10 V then reducing the voltage to 0 V on the first electrode as illustrated in Figure 15.19a. Hence by pulsing the voltages on the electrodes sequentially between high and low levels, charge signals can be made to pass along an array of electrodes. To achieve this, the electrodes are connected in sequence to a set of three-phase drive pulses as illustrated in

Figure 15.19b. Charge signals can then be stored under every third electrode in the array and will be transferred together along the array under the control of the drive pulses. The use of three phases ensures that the charges move in the right direction. By letting the presence or absence of a charge represent digital values of 0 and 1, and by providing amplifiers for injecting and detecting these charges, a very simple and compact type of computer memory device is possible. When a CCD is used as an electronic sensor in a camera, the metal electrodes are overlaid by surfaces that are optically sensitive.

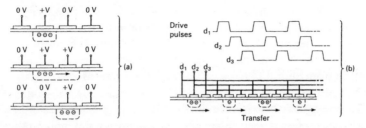

Figure 15.19 How the charge is transferred in a CCD.

A different type of light-sensitive semiconductor array is the CMOS sensor that is a rival to the CCD sensor. They compare as follows:

▶ *The CMOS sensor is more susceptible to electrical noise than the CCD sensor.*
▶ *Since each pixel on a CMOS sensor has several transistors located next to it, its light sensitivity tends to be lower since many of the photons hit the transistors rather than the light-sensitive CMOS.*
▶ *CMOS consumes up to 100 times less power than the CCD sensor.*
▶ *CMOS is cheaper than CCD since it can be made on any standard production line making silicon chips (see Chapter 12).*

Thus CCDs tend to be used in digital cameras if high-quality images are required with good light sensitivity.

15.10 Fibre optics communications

For a long time, light has been used to send messages. Our ancestors lit beacons when invaders threatened. A hand-held mirror, the heliograph, was first used by the ancient Greeks to reflect the Sun's rays and flash coded signals over great distances. And lighthouses and traffic lights use light to warn us of danger. But fibre optics communications is an altogether more sophisticated way of sending messages from one place to another: it makes use of long thin glass fibres along which information is sent as pulses of laser light. Fibre optics communication is now a well-established technology and it holds great promise for the future of telecommunications systems. But why should optical fibres be so superior to conventional copper cables? Well, for one thing, cables made from optical fibres are cheaper, lighter and easier to install than copper cables. Furthermore, they are completely free from electromagnetic interference since data on a light beam cannot be corrupted by electrical machinery, thunderstorms and 'noisy' power lines. Consequently, there is no interference or 'cross-talk' between neighbouring fibres, a quality that also means that signals carried by optical fibres are much less liable to be detected compared with electrical signals in copper cables, so the information is effectively secure from eavesdroppers. Safety, too, is an important reason for using optical fibres since broken fibres are not a fire hazard as the escaping light is harmless. But perhaps the strongest justification for using optical fibres is their potential for carrying considerably more information than copper cables. A glance back at Figure 15.2 shows why. Since light waves have frequencies about 10,000 times higher than the highest frequency radio waves, considerably greater bandwidth is available. Indeed conventional copper cables are hard pressed to keep up with the mounting speed of development in communications and information technology. However, before fibre optics communications could become a reality, two high technology inventions, the laser and low-loss fibre optic cable were needed.

By 'low loss' is meant glass so pure you could see through a 35 km thick block of it as clearly as through a window pane! Such high purity means that information travels through optical fibres for long distances without having to be repeatedly amplified on the way – it is said to have low attenuation. The raw material for such optically clean fibres is a special kind of sand called silica. An optical fibre is a solid rod of silica, finer than a human hair and surprisingly flexible. It is manufactured in the cleanest of atmospheres to ensure that no speck of dust or fingerprint can mar its purity.

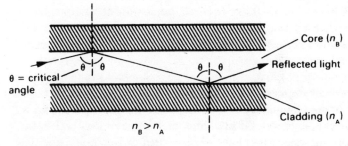

Figure 15.20 How light travels through an optical fibre.

The basic construction of an optical fibre is shown in Figure 15.20. It comprises a glass core totally enclosed by a glass cladding. A plastic coating covers the cladding and core to prevent dust and moisture from reaching the glass core. Light (actually, it is infrared – see below) does not just travel along the core in any old way; it is guided along the core by being reflected back from the outer cladding to the core, bouncing along from side to side of the core. No light is lost as it bounces from the cladding; at each bounce it is all reflected back to the core. What makes this happen is the relative optical properties of the core and the cladding. The cable is designed so that the refractive index of the core is higher than the refractive index of the cladding. This ensures that light meeting the boundary between the core and cladding at an angle greater than a certain 'critical' angle is totally reflected back into the core. This is called total internal reflection. A fibre optics cable is an armoured cable designed both to protect the bundle of fibres from contact with moisture and chemicals, and to strengthen it.

Readers who know some basic optics may recall that a simple equation is used to calculate the critical, or minimum, angle, θ, at which the light must meet the boundary if it is to be totally internally reflected. It is

$$\sin\theta = n_A/n_B$$

where n_B is the refractive index of the glass of the core and n_A is the refractive index of the cladding. Thus, if n_A = 1.43 and n_B = 1.45, $\sin\theta$ = 1.43/1.45 = 0.9862. Therefore, θ = 80°28'. For angles of incidence less than θ, the light passes into the cladding and is lost. Of course, glasses with different refractive indices give different values of critical angle.

Two light sources (or transmitters) are eminently suitable for the job of sending pulses of light down these slender optical fibres: light-emitting diodes (LEDs) and injection-laser diodes (ILDs). Both sources generate light when excited by electricity, and they are the only sources of light capable of being switched on and off fast enough to be modulated by low-power analogue or digital signals. Their physical dimensions are compatible with optical fibres and they have the reliability and long life needed in telecommunications systems. Gallium arsenide (GaAs), gallium aluminium arsenide (GaAlAs) and gallium indium arsenide phosphide (GaInAsP) are the materials used in their construction. They convert electricity into infrared energy efficiently, and the special glass used in fibre optics is more transparent at infrared wavelengths. Infrared LEDs are suitable for use with stepped-index multi mode fibres since they emit a relatively wide beam of light with a fairly large spectral bandwidth. ILD sources radiate a much narrower beam of infrared with a much narrower spectral width. Thus ILD sources are ideal for use with stepped-index mono-mode fibres. Furthermore, ILDs can launch between 0.5 mW to 5 mW of infrared power into a fibre, compared with the smaller 0.05 mW to 0.5 mW for an LED. And ILDs can be modulated at frequencies of over 500 MHz compared with about 50 MHz for LEDs. Gallium aluminium arsenide ILDs and LEDs generate infrared in the 0.8 μm to 0.9 μm range, while gallium indium arsenide phosphide devices generate infrared in the 1.3 μm to 1.6 μm range where attenuation and dispersion by fibres is very low.

A photodiode is generally used to convert the modulated infrared light back into electrical signals at the end of the fibre. The photodiode is reverse-biased so that when it absorbs infrared, a small current flows between its cathode and anode terminals. The current is virtually proportional to the amount of light it absorbs. Photodiodes are generally based on silicon. Infrared emitting diodes and photodiodes are available as spectrally matched pairs: the LED emits maximum infrared radiation at the wavelength to which the photodiode is most sensitive. This wavelength is typically 0.9 μm.

Insight

The ideal world of data transmission is to minimise distance by being able to send limitless amounts of data extremely quickly. Although the latest fibres can transmit 10 terabits (10 million million bits) of information in two minutes: an amount equal to all the 40,000 movies that have ever been produced, to make video conferencing possible with several participants needs one thousand times more than this data rate.

15.11 Pulse code modulation (PCM)

It is not usual to send messages along optical fibres using amplitude modulated infrared. Commercial optical communications systems use digital techniques to produce pulses of infrared coded in some way to carry the information required. There are several ways of producing digitally coded information. One of these, called pulse code modulation (PCM), is used in radio as well as optical communications systems. Indeed, it is the preferred method for communicating with spacecraft and communications satellites.

In pulse-code modulation an analogue signal is sampled at intervals as shown in Figure 15.21 to give a pulse amplitude modulated

(PAM) signal. To ensure that the information can be transmitted and recovered without undue reduction of intelligibility, the frequency at which the analogue signal is sampled has to be at least twice the bandwidth of the signal. Now although the healthy ear responds to signals in the frequency range of 20 to 20 000 Hz, voice is intelligible in the frequency range 300 to 3400 Hz. Therefore a sampling frequency of 8 kHz is used to accommodate voice transmissions. Generally the higher the sampling frequency, the better the copy of the original analogue signal that is recovered at the destination.

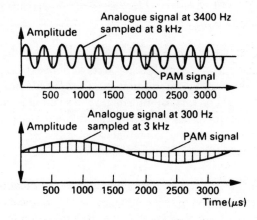

Figure 15.21 The sampling of an analogue signal so as to produce a pulse amplitude modulated signal.

The amplitude of each sample is measured using a scale of levels. This measurement is then encoded into a binary equivalent using an analogue-to-digital converter (Chapter 14), and this binary equivalent is known as a pulse code modulated signal. The technique is shown in Figure 15.22. Here a three-bit binary code is used to represent eight possible (0 to 7) levels of the sample taken at equal intervals of time, t_1 to t_2. The binary equivalent of each sample is then transmitted as a pulse waveform. At the destination a digital-to-analogue converter decodes the digital signal back into an analogue one. It is the number of bits per sample, or its resolution, which determines the accuracy of the magnitude of the analogue signal at the point the sample is made. Thus, for an eight-bit binary number, 256 levels represent each sample.

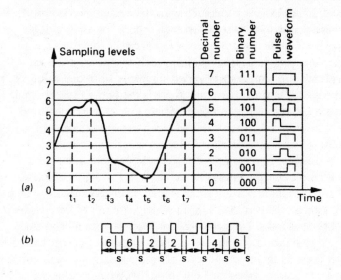

Figure 15.22 (a) Assigning a 3-bit binary code to each sample; (b) the transmitted pulse waveform.

There are several advantages of using pulse code modulation. First, it is a powerful method of overcoming noisy environments since it is only the presence or absence of a pulse that is needed to construct the original signal, not its shape. In this way it is possible to glean information from extremely weak signals received from spacecraft transmitting information from the edge of the Solar System. Second, in a long-distance communications link, the signal is apt to lose energy and be lost and, for Earth-based systems, the power of the signal is regenerated en route using repeaters. However, in an analogue signal, not only the signal but also the noise is amplified. In the case of a digital signal, the signal is regenerated as a pure, clean signal since only the on/off states of the pulse are identified. Third, once a signal has been digitized it is compatible with other digital signals. For these reasons it is not surprising that digital signal processing is rapidly overtaking the older analogue systems, nowhere more so than in the broadcasting of radio and TV programmes.

Insight

In this book I have often compared the characteristics of the analogue world with the digital world. Indeed, Chapters 14 and 15 are largely concerned with the circuits needed for these two worlds to speak to each other. The assumption we have made is that our senses experience the everyday analogue world as if its qualities such as colour, pressure, volume and temperature varied smoothly from one value to the next. On the other hand, there is the human constructed digital world of the computer where qualities are measured in patterns of 1s and 0s. However, if we look more deeply into the analogue world events and processes seem less continuous. For example, look closely at the minute hand of a mechanical analogue watch and you will notice that the hand moves forward in small steps. This is because the watch has an escapement mechanism which determines the passage of time as ticks. Furthermore, quantum physics tells us that light radiation is composed of small packets of energy called photons. Is reality at the level of atoms a digital world and does it become smoothed out in our everyday world?

TEST YOURSELF

1 *Draw a labelled diagram that shows the essential elements of a communications system.*

2 *Make a list of the following types of electromagnetic radiation in descending order of frequency:*
   ultraviolet radiation; gamma rays; medium frequency radio waves, infrared radiation; X-rays

3 *Match each of the terms in the left-hand column of the following table with an appropriate unit in the right-hand column:*

1 frequency	A seconds
2 wavelength	B metres per second
3 amplitude	C metres
4 period	D millivolts
5 velocity	E hertz

4 *If a radio station broadcasts on a frequency of 100 MHz, what would be the length of a dipole designed to receive this broadcast?*

5 *When a carrier wave is frequency modulated, is the carrier frequency varied in proportion to the*
   (a) *amplitude of the modulating signal, or*
   (b) *frequency of the modulating signal?*

6 *Is a radio receiving aerial designed to convert*
   (a) *air pressure into electrical energy, or*
   (b) *electrical energy into electromagnetic energy, or*
   (c) *electromagnetic energy into electrical energy?*

7 The radio medium waveband extends from 500 kHz to 1.5 MHz. If a radio station occupies a bandwidth of 10 kHz:
  (a) How many channels can be fitted into the waveband?
  (b) What will be the highest frequency of sound that is transmitted from this station?

8 Describe how fibre optics enhances telecommunications.

9 What are the main advantages of digital radio and TV communications compared with analogue communications?

10 What is meant by pulse code modulation? If the sample rate is 10 kHz, what is the time delay before the next sample has to be transmitted?

16

....

Projects

In this chapter you will learn how to assemble working circuits using a breadboard assembly system that does not involve soldering components together.

16.1 Introduction

The study and application of electronics is such a creative activity that some readers may be encouraged to construct working circuits that make use of some of the concepts introduced in the previous chapters. To this end, I have drawn up seven introductory projects which can be easily assembled at an affordable cost. The projects are:

Project 1: *Circuit Tester*
Project 2: *Dark Switch*
Project 3: *Games Timer*
Project 4: *Frost Alert*
Project 5: *Rain Alarm*
Project 6: *Simple Die*
Project 7: *Simple AM Radio*

A version of these projects first appeared in the journal *Everyday Practical Electronics* (EPE) during 2008/9 – see the section Taking it Further. If you want to extend your practical skills, this journal also included more sophisticated projects that I designed for breadboard assembly. These include a Bat Detector, Ultrasonic Remote Control and a Lightning Detector.

A simple means of circuit assembly is to use a short length of terminal strip as in Project 1, (see Figure 16.3 and 16.4), which enables components to be connected together using a small screwdriver to tighten screws that grip component leads. The rest of the projects are assembled on a 'breadboard', a photo example of which is shown in Project 3. The particular breadboard I have used is called 'Protobloc' meaning a 'block' or board which is used to assemble a working circuit without needing to solder components together. A circuit assembled on Protobloc lets you add and replace components and to change connections as necessary. Also it means that the components can be reused in other projects.

The Protobloc I have used has 400 small holes, each one allowing a component lead to be inserted and gripped by double leaf spring contacts underneath each hole. Link wires are used to bridge across these sockets enabling components to be interconnected. These sockets are arranged in a grid pattern, which are precisely 0.1 inch (2.54 mm) apart. This separation is to accommodate integrated circuits which have terminal pins spaced apart by exactly this amount. The holes of each line of the two lines of holes at the 'top' are all joined together as are the ones at the 'bottom' of Protobloc. It is normal practice to connect the positive and negative of a power supply to these rows. A 9V battery, e.g. a PP3 or PP6, is used for all the projects. The 'vertical' lines of holes are linked together in groups of five and there is no link across the central channel of the breadboard. This enables each pin of an integrated circuit to be accessed by connections made to the remaining four sockets in the 'vertical' groups of socket on each side of the integrated circuit.

When using Protobloc, you should not try to mimic the layout of components in the circuit diagram. Instead you should concentrate on the connections between the components in achieving the neatest possible arrangement of components. In the projects I have done this to enable you to follow the interconnections easily. In some cases this has meant using more links than is necessary and you may well find a more efficient use of link wires. However, I advise you not to modify the layouts as shown until you become more experienced with using Protobloc to evaluate your own circuit designs.

To ensure trouble-free assembly and a successful working project when using Protobloc, you should try and follow these six important rules:

1 Always use single-core plastic-coated wire of 0.6 mm diameter for wire links, not thicker. The ends of the wire should be stripped of plastic for about 8 mm. The use of thicker wire can permanently damage the springy sockets underneath each hole. Wire links already cut to length with bare ends already bent at right angles are available from suppliers.

2 Never use stranded wire as it can fray and catch in the sockets, or a strand can break off and cause unwanted connections below the surface of the Protobloc.

3 It is very important to make sure that bared ends of link wires and component leads are straight before inserting them into the Protobloc. Kinks in the wire will catch in the springy clip below the socket and damage it if you have to tug the wire to release it from the holes. You should use snipe-nose pliers (see below) to straighten leads that are kinked.

4 Make sure that the arrangement of components and wire links are tidy, with components fitting snugly close to the surface of the Protobloc. This usually means providing more link wires than is perhaps necessary so as to avoid having wires going every which way across the Protobloc. Your finished assembly should be a tidy work of art!

5 Never connect the battery leads to the top and bottom rails of the Protobloc until you have carefully checked that all the connections correspond to those on the circuit diagram.

6 Some components such as switches and relays do not have appropriate wire leads for insertion into the Protobloc. If you have access to a soldering iron, solder short lengths of 0.6 mm diameter plastic coated wire to the terminals of the component. If you don't have a soldering iron, and don't know anyone who has, then resort to the less satisfactory solution of using leads with crocodile clips on the ends. Some leads such as the battery connectors can be prepared by connecting each of them to a single section of terminal strip and anchoring short lengths of 0.6 mm diameter wire to the other side of the block as shown in Figure 16.1.

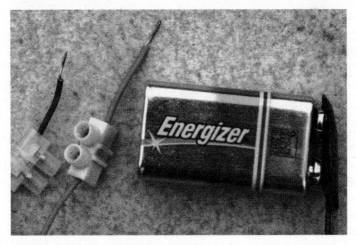

Figure 16.1 How to provide battery leads for Protobloc assembly.

TOOLS

Figure 16.2 shows the three basic tools needed for assembling circuits on Protobloc. The screwdriver should be a 2 mm flat blade type that will be useful for adjusting preset variable resistors, for example. The snipe-nose pliers are ideal for picking up small components and for straightening and inserting component leads into Protobloc. The wire cutters/strippers have an obvious use. These tools and the components needed for the projects can be found from suppliers that advertise in *EPE*.

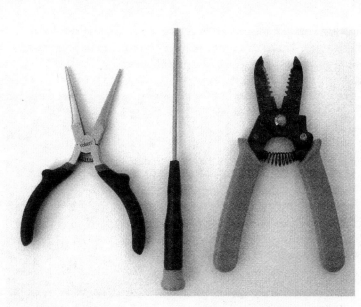

Figure 16.2 Tools required for the projects.

16.2 Project 1: Circuit Tester

You will find that the simple project shown in Figure 16.3 is
a handy way of checking the electrical properties of resistors,
capacitors, transistors, light-dependent resistors and thermistors
that have featured in the chapters of this book. To use the Circuit
Tester you connect the crocodile clips to the leads of a component
and keep an eye on the brightness of the light-emitting diode
(LED_1) which indicates the electrical resistance between the clips.
It works by trying to pass a small current through the component
being tested, the light-emitting diode shining brightly, dimly or
not at all according to the resistance of the component which
is connected in series with it. Materials having a high resistance
such as plastics and paper will not allow much current to flow
through them so LED_1 will not light, whereas materials with a low
resistance such as metals will allow current to flow easily through
them as shown by LED_1 shining brightly. Note that the current

flowing between the crocodile clips is direct current provided by a 9V battery. This d.c. current enables you to test the polarity of a number of components, as described below.

Figure 16.3 The circuit tester ready to use.

COMPONENTS AND MATERIALS

► *Fixed value resistor, R1: value 220 ohms, ¼ W carbon film. Its value is colour coded, the bands being red, red, brown, and fourth silver band indicating the 10% tolerance of the resistor – see Chapter 5.*
► *Two 10 mm red light-emitting diodes one used as, LED$_1$ and the other spare for testing purposes.These large LEDs are easily seen. The anode lead is the longer lead – see Chapter 11.*
► *9V PP3 battery clip and battery.*
► *Test leads terminated at one end with a crocodile clip: one red and one black. Remove the clip from one end of a black and red lead.*
► *Two 6-way lengths of terminal strip, one for the project and the other spare. Or one 12-way block cut in two. One small project box, e.g. 59 × 87 × 31 mm.*

- ▶ *Short lengths of 0.6 mm diameter insulated wire for making links on the terminal strip. I suggest you buy a jump wire kit that contains a variety of lengths of plastic-covered wire of the right diameter for use in terminal block and with breadboards.*
- ▶ *You will also need a small screwdriver, wire stripper and cutter, a set of small drills and a hand drill or electric drill.*

ASSEMBLY

Follow the wiring on terminal strip as shown in Figure 16.4a making sure that LED$_1$ is connected so that its anode lead is in the position shown on the terminal strip. The other lead is known as the cathode. Usually, the anode is the slightly longer lead. The resistor, R_1, is connected in series with LED$_1$ and is required to ensure that too much current does not flow through it when the crocodile clips are connected together.

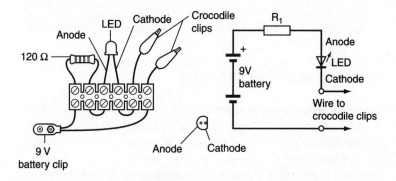

Figure 16.4(a) Assembly of the circuit tester using a length of terminal strip (b) circuit diagram.

Glue the Circuit Tester to the base of the box. Drill a hole to allow LED$_1$ to poke through the lid when it is closed. Do not forget to make a knot in the crocodile lead cables inside the box to stop the leads pulling on the terminal strip. Once the project is assembled, clip the two crocodile clips together and LED$_1$ should light brightly showing negligible electrical resistance between the leads. Follow the simple activities described below to find out how to use the Circuit Tester but first note the following precautions.

When the Circuit Tester is not in use, clip the leads to a piece of rigid insulating material such as cardboard or plastic to avoid them accidentally touching and draining the battery. This avoids having an on-off switch.

Never use the Circuit Tester to test mains-operated equipment whether that equipment is switched on or switched off.

USING THE CIRCUIT TESTER

The following activities will give you a feel for ways of using this simple tester in the projects that follow. BUT, AGAIN, DO NOT TEST ANYTHING THAT IS CONNECTED TO THE MAINS POWER SUPPLY!

Activity 1 Fuse tester
You have a ready-made fuse tester but remember never to test any fuse or any device that is connected to the mains electricity supply. Press the crocodile clips across the metal ends of the fuse – if the LED shines brightly, the fuse is good.

Activity 2 Random resistance tester
Test different materials and objects for their ability to conduct electricity by clipping the crocodile clips across two points on them. For example: the lead of a pencil; brass and silver objects around your room; the resistance of your upper body between your hands (moisten your fingers with saliva before gripping the clips – the LED should just glimmer); a pendant, the stem of a plant, etc. Remember that low electrical resistance between the clips causes the LED to glow more brightly than does higher resistance. Draw up a list recording high, medium and low resistance of these items.

Activity 3 Salty water conducts
Fill a cup with tap water and then dip the metal ends of the crocodile clips just under the surface separated by a couple of

centimetres. The LED should barely glow, indicating that tap water does not have a particularly low or high resistance. Now add a sprinkling of ordinary table salt to the water and stir it in. What effect does this have on the brightness of the LED? How has the salt changed the resistance of the water?

Next, moisten a dish cloth or sponge with the salty tap water. Clip one crocodile clip to the edge of the dish cloth and touch the other crocodile clip to different parts of the cloth. Watch the LED fade and brighten as you do this because the resistance between the clips varies depending on how much moisture exists between the clips and their distance apart. How might you use the Circuit Tester to find out whether a plant pot needs watering? Try it!

Activity 4 Light sensor
You need a light-dependent resistor (LDR) for this activity, a widely used semiconductor that changes its electrical resistance with the amount of light falling on it – see Chapter 5.

Clip the crocodile clips to the two leads. It doesn't matter which crocodile clip goes to which lead. Allow more or less light to reach the sensitive face of the LDR and at the same time watch the brightness of the LED on the Circuit Tester. What can you say about the resistance of the LDR when it is brightly illuminated compared with when it is in shadow? Change over the connections of the crocodile clips and you'll notice no difference in the way light changes the electrical resistance of the LDR – it is said to be non-polarised.

Activity 5 Temperature sensor
You need a thermistor for this activity. Like, the LDR, the thermistor is a widely used semiconductor: it changes its electrical resistance as it warms up and cools down – see Chapter 5.
Clip the crocodile clips to the two leads. It doesn't matter which crocodile clip goes to which lead. Put some warm water in a cup and dip the thermistor just under the surface at the same time watch the brightness of the LED on the Circuit Tester. Next dip the thermistor into cold water. What can you say about the resistance of the thermistor when it is warm compared with when it is cool?

Change over the connections of the crocodile clips and you'll notice no difference in the way temperature changes the electrical resistance of the thermistor – it is said to be non-polarised.

Activity 6 One-way traffic

Some electronic components have the ability to stop electricity from flowing through them in one direction and allowing current to flow in the opposite direction. They are generally known as diodes – see Chapter 7. Let's examine a light-emitting diode (LED) such as the one you used for making the Circuit Tester. First connect the crocodile leads across the LED leads any way round. Does the LED on the Circuit Tester glow? If it doesn't, change over the connections and the LED on the Circuit Tester should light and the red crocodile clip identifies the anode terminal of the LED. The LED is said to be 'forward biased' by the positive voltage on the red croc clip enabling current to flow through the LED being tested. If the LED on the Circuit Tester does not glow, it is because the red croc clip is connected to the cathode terminal of the test LED which is said to be reverse biased and no current can flow through it. Note that if the LED in the Circuit Tester does not light for both positions of the crocodile clips, the LED under test is probably damaged. The test that you have just done works equally well for finding the anode terminal of a rectifying diode.

Activity 7 Transistor tester

Transistors of the npn and pnp variety are called bipolar transistors since they are made from n-type and p-type semiconductors through which electricity flows mostly by negatively charged electrons or mostly by positively charged holes, respectively. The circuit symbols for these two transistors are shown overleaf and identify the emitter (E), base (B) and collector (C) terminals. Thus npn and pnp transistors can be considered as comprising two diodes connected as shown in Figure 16.5. Once you have this structure in mind, you can identify whether a transistor is pnp or npn using the Circuit Tester as you did for the LED above.

All you need to do is to find the colour of the crocodile clip that, when attached to one of the transistor terminals, causes the LED on the Circuit Tester to light up when the black clip is attached to each of the remaining two terminals in turn. This would then be an npn transistor and the red lead identifies the base terminal. If it is a pnp transistor, the black lead identifies the base connection if the red lead causes the LED light up when it is attached to each of the remaining two terminals. This simple test does not identify which of the remaining transistor terminals is collector or emitter.

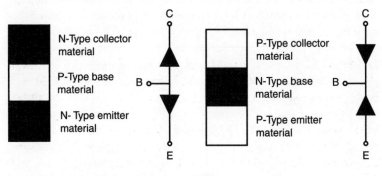

Figure 16.5 Representation of bipolar transistors as two coupled diodes.

16.3 Project 2: Dark/Light Switch

The Dark/Light Switch circuit shown in Figure 16.6 uses two common and cheap general purpose bipolar transistors in a semiconductor switch that responds to changes of light intensity falling on the light-dependent resistor, LDR_1. The two transistors are coupled together in an arrangement known as a Darlington pair which was discussed in Chapter 8. The circuit can be regarded as made up of three interlinked building blocks. Building block 1 is the input to the system, the dark sensing function, which is based on VR_1 and LDR_1. Together these act as a voltage divider. The voltage at the junction of these two components rises and falls according to the amount of light falling on the LDR. The voltage divider is followed by building block 2 based on Tr_1 and Tr_2 acting as a sensitive switch and providing sufficient current to energize building block 3 comprising LED_1 and the relay connected in parallel with it. As the light level decreases and exceeds a critical point, the relay energizes and its normally open (NO) switch contacts close. These contacts can then be used to switch on motors or lamps, for example. The light intensity when switching occurs is determined by the setting of VR_1. An alternative use is made of the circuit by interchanging LDR_1 and VR_1, which causes the relay to energize when the ambient light level rises above a particular intensity. The circuit shows the dark-operated version so that the relay can switch on a motor, a buzzer or a lamp when the light level falls, such as at dusk.

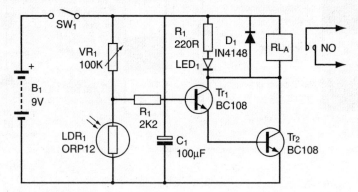

Figure 16.6 Dark/Light switch circuit.

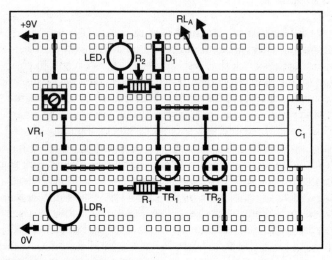

Figure 16.7 Breadboard layout of the Dark/Light Switch.

COMPONENTS AND MATERIALS

▶ *Npn transistors, Tr₁ and Tr₂: both BC108. Any general purpose types will do instead. For emitter, base and collector terminal connections, see Section 8.5.*

▶ *Light-dependent resistor, LDR₁: ORP12 type or similar – see Section 5.6. Fabricate a paper tube to fit round the LDR to provide better directional light sensing.*

▶ *Variable resistor, VR₁: 100 kΩ for the dark-operated version. A higher value may be necessary for the light-operated version. The breadboard layout shows a miniature preset type – see Section 5.3.*

▶ *Fixed-value resistors: R₁ and R₂: 2.2 kΩ and 220 Ω, respectively, each ¼ watt type. For colour coding see Section 5.4.*

▶ *Light emitting diode, LED₁: red, say, as an indicator of an energized relay.*

▶ *Rectifying diode, D₁: e.g. 1N4148 type since its leads are thin enough for insertion into the Protobloc – see Chapter 7. It is used to short-circuit the back e.m.f. generated when the relay de-energizes, which could damage the transistors.*

▶ *Relay, RL_A: 6V type, single pole, changeover – see Section 3.5.*

▶ *Electrolytic capacitor, C₁ with a value of 100 μF or larger: this stabilizes the power supply voltage – see Chapter 6.*

▶ *Protobloc and wire links.*

16.4 Project 3: Games Timer

This project produces an audio alarm after a preset time delay. In the circuit of Figure 16.8, push switch SW_1 is pressed momentarily to start the timing. The circuit essentially comprises one building block based on an integrated circuit 555 timer wired as a monostable as explained in Chapter 6. Variable resistor, $(VR_1 + R_2)$ and capacitor C_2 determine the time delay, which is between about 10 seconds and 3 minutes using the component values shown. At the end of the timing period, the voltage at pin 3 falls and the buzzer sounds. When the circuit is switched on by means of SW_2, the buzzer sounds at first but stops once the circuit is triggered to come on again after the time delay. When LED_1 is lit, it shows that the circuit is ready to be triggered or the timing has ended, while LED_2 lights during the timing period. You can increase the values of VR_1 and C_2 to provide longer time delays. Capacitor, C_1, offers a short circuit when SW_1 is pressed, momentarily causing the trigger pin 2 of IC_1 to be connected to 0V to start the timing. Resistor R_4 connected across capacitor C_1 ensures that the capacitor discharges once it becomes momentarily charged by pressing SW_1, while resistor, R_1, ensures that the voltage on the trigger pin is normally at the positive supply voltage. A dial attached to VR_1 may be calibrated in seconds and minutes as required.

COMPONENTS AND MATERIALS

- ▶ *Integrated circuit, IC_1: 555 timer – see Section 6.7.*
- ▶ *Variable resistor, VR_1: 1 MΩ – see Section 5.3.*
- ▶ *Fixed-value resistors, R_1, R_2, R_3, R_4 and R_5: 10 kΩ, 10 kΩ, 220 Ω, 470 kΩ, and 220 Ω, respectively, all ¼ watt – see colour codes in Section 5.4.*
- ▶ *Capacitors C_1 and C_2: 10 µF and 470 µF, respectively, both electrolytic and 16 V or higher working voltage. For C2, you could also use values between 100 µF and 1000 µF – see Section 6.4. Note that capacitors C_1 and C_2 are polarized and require connecting as shown.*
- ▶ *Two light emitting diode, LED_1 and LED_2: red and green. If you are unsure, use the Circuit Tester to distinguish between the anode and cathode leads of each one.*
- ▶ *Buzzer, BZ_1: miniature 6 V type.*
- ▶ *Switch, SW_1: push-to-make, release-to-break.*
- ▶ *9V battery, B_1, and clip.*

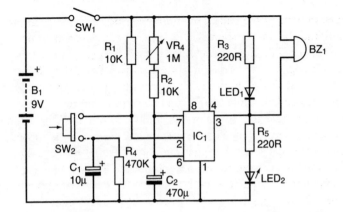

Figure 16.8 Games timer circuit.

Figure 16.9 Photo of games timer.

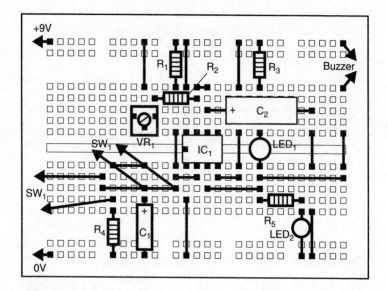

Figure 16.10 Breadboard layout of the games timer.

16.5 Project 4: Rain Check

This little project shown in Figure 16.11 sounds an audio alarm when a simple sensor becomes wet with rain. Building block 1 is the sensor, S_1; building block 2 is the switch based on Thy_1; and building block 3 is the output which is the visual signal provided by LED_1 and the audible signal provided by BZ_1. The sensor, S_1, is made as described in the notes below. When the gaps between the conducting strips become wet, the electrical resistance between the different conducting parts of the sensor falls. This causes the voltage to rise on the gate terminal switching on the thyristor as the voltage on the anode terminal falls. Current then flows through the buzzer, BZ_1 which continues to sound until it is reset by unplugging and reinserting one of the battery leads. You will find that bridging the strips with your finger will switch on the alarm, and there you have a touch switch! Please read the above rules for assembling circuits on breadboard before tackling this project. Once the circuit is assembled and checked carefully, the power supply can be connected.

COMPONENTS AND MATERIALS

▶ *Thyristor, Thy₁: type 2N5060 in a TO92 style package – see Section 7.6.*
▶ *Resistors, R₁ and R₂: values 2.2 kΩ and 270 Ω, respectively both ¼ watt, carbon film – for colour code see Section 5.4.*
▶ *Buzzer, BZ₁: operating voltage 6 V.*
▶ *Relay, RLₐ: low voltage 5 V or 6 V type, single-pole changeover.*
▶ *Capacitor, C₁: axial electrolytic, 220 µF, 25 V – see Section 6.4.*
▶ *Switch, SW₁: single-pole single-throw (optional).*
▶ *Sensor, S₁: Make it from a small piece of stripboard, linking alternate tracks together with short lengths of wire to provide an interleaved set of tracks.*
▶ *9 V battery, B₁, and connecting leads.*
▶ *Protobloc and wire links.*

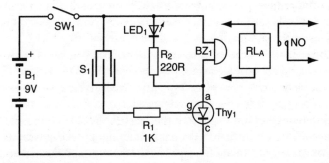

Figure 16.11 Rain check circuit.

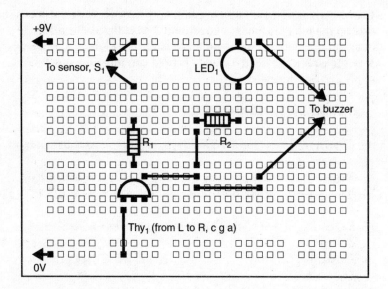

Figure 16.12 Breadboard layout of the rain check.

16.6 Project 5: Frost Alert

This project produces an intermittent audio alarm when the air temperature reaches the freezing point of water, which is 0°C in normal conditions of atmospheric pressure. The temperature sensing device is a thermistor, Th_1, the resistance of which decreases with increasing temperature – see Section 5.6. The circuit shown in Figure 16.13 comprises three main building blocks. The first building block acts as a temperature-sensitive electronic switch based on an integrated circuit operational amplifier, IC_1, which is activated when the thermistor reaches 0°C – see Chapter 13. Once activated, it energises the second building block comprising two coupled 555 timers, IC_2 and IC_3 that are interconnected as two coupled astables – see Section 6.8. The varying signal from the second building block is amplified by the third building block comprising the npn transistor so the small speaker produces a loud pulsing when the thermistor's temperature falls to 0°C.

In order to understand in more detail how the circuit works, focus first on the building block centred on IC_1 which is connected as an electronic switch. Notice that the voltage divider action of resistors R_1 and R_2 provides a fixed voltage on the non-inverting pin 3 of IC_1. This voltage is compared with the variable voltage on its inverting pin 2, which is determined by the resistance of the thermistor, Th_1, and of the setting of VR_1. Once this variable voltage falls below the voltage on pin 3, the output voltage of IC_1 rises sharply to near the positive supply voltage. Now this positive voltage is sensed by pin 4 of IC_2 which is wired as a low frequency astable causing it to oscillate at a frequency set by the values of C_1, R_5 and R_7. This frequency is about 1 Hz and is monitored by the flashing of LED_1.

Each time the output of IC_2 goes HIGH (that is, positive, which is shown by LED_1 glowing red) it enables the second astable based on IC_3 via its pin 4. When the output voltage of IC_2 falls to 0V, it switches IC_3 off. Now the frequency of the astable that is centred on IC_3 is 100 times higher than that based on IC_2, so that the two astables provide a continuous 'beep, beep,...' sound from LS_1.

The circuit is adjusted so that it responds to a temperature of 0°C by immersing the thermistor in melting ice (to make sure it is at 0°C) and adjusting the variable resistor, $VR1$, so that the alarm is just triggered.

COMPONENTS AND MATERIALS

▸ *Thermistor, Th_1: disc-shaped having a resistance of 5 kΩ at 25°C; e.g. NTC type TTC502A – see Section 5.6.*
▸ *Integrated circuits, IC_1, IC_2 and IC_3: CMOS op amp type 7611, and two 555 timers, respectively – see Chapter 12.*
▸ *Fixed-value resistors, R_1 to R_8: 47 kΩ, 47 kΩ, 1 MΩ, 220 Ω, 47 kΩ, 47 kΩ, 47 kΩ, and 47 kΩ, respectively, all ¼ W – for colour codes see Section 5.4.*
▸ *Variable resistor, VR_1: 50 kΩ.*
▸ *Capacitors C_1 to C_3: 10 µF (electrolytic), 10 nF (polyester) and 10 µF (electrolytic), respectively. Note that capacitors C_1 and C_3 are polarized as indicated by the '+' sign at one end.*

- *Light-emitting diode, LED₁: If unsure, use the Circuit Tester described above to identify the anode lead for LED₁.*
- *Loudspeaker, LS₁: 8 or 16 Ω.*
- *9 V battery, B₁, and connecting lead.*
- *Protobloc and wire links.*

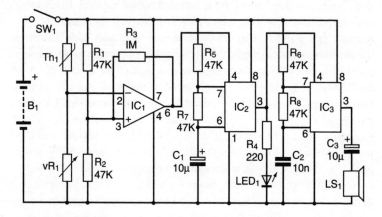

Figure 16.13 Frost alert circuit.

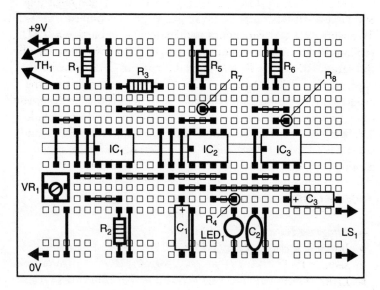

Figure 16.14 Breadboard layout of the Frost Alert.

324

16.7 Project 6: Simple Die

The circuit shown in Figure 16.15 makes an interesting electronic alternative to the traditional little cube, the die, with its collection of dots on its six faces. The circuit is based on the CMOS 4017 decade counter, IC_2, which is described in Chapter 10.3. This has ten outputs that go HIGH (that is positive) in turn with each successive 'clock' pulse produced by a 555 timer, IC_1, operating as an astable – see Chapter 6.8. This is achieved by connecting the output pin 3 of the 555 timer to the clock input pin 14 of the 4017 decade counter. However, Instead of allowing the 4017 to cycle through all of its ten states, the reset pin 15 is connected to the Q_6 output (pin 5) so that the 4017 cycles through states Q_0 to Q_5, only, resetting to Q_0 as pin 5 goes logic HIGH. When each one of the six outputs Q_0 to Q_5 goes logic HIGH, the associated LED lights.

The 'roll' of the die is accomplished by pressing the push-to-make switch, SW_1, to generate the clock pulses. These are generated at a rapid rate and all of the LEDs appear to be lit. Upon releasing SW_2, one of the six outputs stays HIGH and the coupled LED is lit. To ensure that a player cannot anticipate when a particular LED remains lit, the astable operates at a high frequency determined by the values of resistors R_1 and R_2 and capacitor C_1. So when SW_1 is pressed, all the LEDs light in rapid succession, too fast for the eye to see the lighting of individual LEDs. Whilst this arrangement does not mimic the dots on the six faces of a die, it does provide an effective but simple electronic solution; it takes the effort (and some would say the fun!) out of rolling a traditional die and picking it up off the floor!

COMPONENTS AND MATERIALS

▶ *Integrated circuit, IC_1: 555 timer – see Section 6.8.*
▶ *Integrated circuit, IC_2: CMOS 4017 decade counter – see Section 10.3.*
▶ *Light-emitting diodes, LED_1 to LED_6: any colour. Use the Circuit Tester to confirm or find the anode lead of each of the six LEDs.*

- *Push switch, SW₁: push-to-make, release-to-break.*
- *Fixed-value resistors, R₁ to R₃: 10 kΩ, 10 kΩ, and 100 kΩ, respectively, all ¼ watt.*
- *Capacitor, C1: 100 nF (0.1 µF), polyester type.*
- *Protobloc and wire links.*
- *9 V battery and clips.*

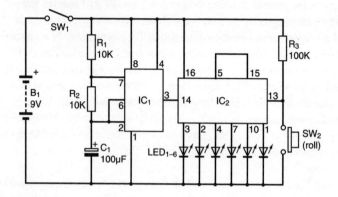

Figure 16.15 Simple die circuit

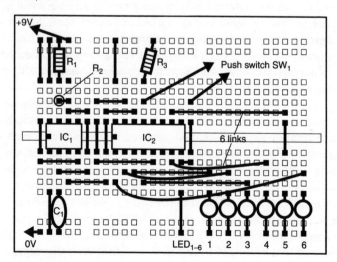

Figure 16.16 Breadboard layout of the simple die.

326

16.8 Project 7: Simple AM Radio

Radios are common enough and cheap but why not make your own radio from a few readily available components? For example, the design shown in Figure 16.17 is uncomplicated thanks to the two integrated circuits, IC_1 and IC_2 and it will receive radio stations transmitting amplitude modulated (AM) radio signals and delivering your selection to a small loudspeaker. The central component for the first building block is the integrated circuit, IC_1, a 'radio on a chip'. Essentially, its purpose is to amplify the small voltage generated across the tuned circuit comprising the LC combination; that is, the variable tuning capacitor, VC_1, in parallel with inductor, L_1. When the radio is 'tuned in' to a station by adjusting VC_1, a small alternating voltage is generated cross the tuned circuit which is processed by the rest of the circuit to extract the information carried by this carrier wave. IC_1 is a 3-pin transistor-like component that requires a 1.5 V supply voltage and this is provided by transistor Tr_1, resistor R_1 and variable resistor, VR_1. Adjustment of this variable resistor alters the supply voltage and determines the sensitivity of the circuit. Capacitor C_3 couples the signal from the radio frequency amplifier and detector stage to the second building block, a low-power audio amplifier based on IC_2 so as to power a small loudspeaker. Variable resistor, VR_2, acts as a volume control.

To set up the circuit, turn VR_2 to maximum volume which in this breadboard layout is fully anticlockwise. Then adjust VR_1 until the radio just stops oscillating. This is the position of maximum sensitivity. VR_2 can then be adjusted to provide a suitable sound volume. Tuning is done with the variable capacitor, VC_1. Note that the receiver is directional so the coil should be rotated for best reception of a particular station. To make the coil, L_1, you will need to wind about 30 turns of enamelled copper wire on the ferrite rod. The tuned circuit comprising L_1 and VC_1 should be connected up using a two-way section of terminal block as shown and the two common 0.6 mm leads plugged into the breadboard.

To enhance the bass response of the speaker, place the speaker in the top of a cup or beaker to provide a sounding box.

COMPONENTS AND MATERIALS

▶ Integrated circuit radio, IC_1: A7642 – see suppliers in Taking it Further.

▶ Npn transistor, Tr_1: BC108. Any general-purpose type will do instead. For emitter, base and collector terminal connections, see Section 8.5.

▶ Audio amplifier, IC_2: 386 type – see suppliers in Taking it Further.

▶ On-off switch, SW_1: single-pole, single-throw.

▶ Ferrite rod: 50 mm length of 10 mm diameter.

▶ Variable (tuning) capacitor, VC_1: value between about 100 pF and 500 pF when adjusted to its maximum value.

▶ Fixed-value polyester capacitors, C_1 and C_3: 10 nF and 100 nF, respectively.

▶ Fixed value electrolytic capacitors, C_2, C_4, C_5 and C_6: 10 μF, 10 μF, 100 μF and 100 μF, respectively.

▶ Fixed-value resistors, R_1, R_2, and R_3: 47 kΩ, 47 kΩ and 1 kΩ respectively, all ¼ watt - for colour coding see Chapter 5.4.

▶ Variable resistors, VR_1 and VR_2: 100 kΩ and 10 kΩ, respectively.

▶ Miniature loudspeaker, LS_1, 8 Ω or 16 Ω resistance.

▶ Coil on ferrite rod, L_1: 30 turns of 0.2 mm enamel-covered copper wire, or pre-wound medium wave aerial coil.

▶ Enamelled copper wire: 0.2 mm diameter and 1 m long.

▶ Knife or sandpaper for scraping off the enamel from the ends of the wire.

▶ 9 V battery and connector.

▶ Protobloc and wire links.

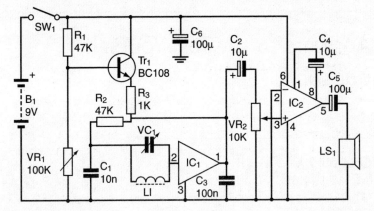

Figure 16.17 Simple AM radio circuit.

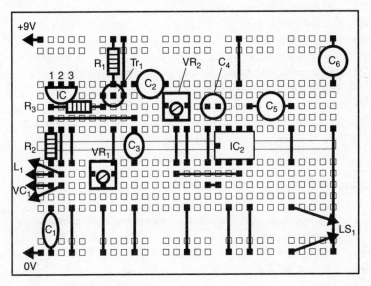

Figure 16.18 Breadboard layout of the simple AM radio.

Answers to the Test Yourself questions

Chapter 1

Q1 *CD player, toaster, security alarm, ...*
Q2 *Emailing, video games, photography*
Q3 *(b) An instrument showing ...*
Q4 *A central processor unit (CPU)*
Q5 *See Section 1.5*
Q6 *It's for you to do!*
Q7 *(b) remains above ...*
Q8 *monitoring heart rate, storing patient medical history, communication between nurses*
Q9 *E.g. weapons development*
Q10 *E.g. chat rooms*

Chapter 2

Q1 *Insulators: plastic and glass. Conductors: copper and aluminium*
Q2 *electrons, protons and neutrons*
Q3 *See Section 2.2*
Q4 *True*
Q5 *Silicon*
Q6 *See Section 2.4*
Q7 *See Section 2.5*
Q8 *See Section 2.5*
Q9 *E.g. www.siliconvalley-usa.com*
Q10 *Hydrogen*

Chapter 3

Q1 *See Section 3.2*
Q2 *Coulombs; a current of one ampere is a charge of one coulomb per second*
Q3 *40 coulombs*
Q4 *3 Ω*
Q5 *See Section 3.5*
Q6 *(b)*
Q7 *Draw a parallel circuit*
Q8 *Draw a series circuit*
Q9 *The answers are the same as for Q8*
Q10 *See Section 3.7*

Chapter 4

Q1 *d.c.*
Q2 *a.c.*
Q3 *True*
Q4 *Electrical current*
Q5 *parallel*
Q6 *low*
Q7 *false*
Q8 *a stream of electrons*
Q9 *4 V*
Q10 *One period is 8 ms. Therefore, the frequency is 1/8 ms = 125 Hz*

Chapter 5

Q1 *electromotive force*
Q2 *the same at all points*
Q3 *3.9 kΩ, 10%*
Q4 *3 kΩ*

Q5   *5.6 kΩ*
Q6   *true*
Q7   *1360 Ω*
Q8   *See Section 5.6*
Q9   *See Section 5.8*
Q10 *thermostat, temperature measurement, fire alert*

Chapter 6

Q1   *Charge large value capacitor (e.g. 10 000 μF) with a 9 V battery and then connect the capacitor to a 6 V lamp.*
Q2   *20 μF*
Q3   *147 nF or 0.147 μF*
Q4   *10^{-3} coulombs*
Q5   *5 s (approx). See Section 6.6.*
Q6   *See Section 6.6*
Q7   *See Sections 6.7 and 6.8*
Q8   *Timer (monostable); flashing indicator (astable)*
Q9   *This is an astable. Use equations for t_1 and t_2 in Section 6.8.*
Q10 *Use equation $T = 1.1 \times C_1 \times R_1$. A monostable.*

Chapter 7

Q1   *True*
Q2   *Forward biased*
Q3   *Figure 7.20a; lamp L1: Figure 7.20b L1*
Q4   *See Figure 7.5*
Q5   *False*
Q6   *See Section 7.3*
Q7   *The diagram should show the anode connected to the positive terminal*
Q8   *Figures 7.5 and 7.10 should be used*
Q9   *See Section 7.6*
Q10 *See Section 7.6*

Chapter 8

Q1 *See Section 8.4*
Q2 *10, 000; 40 dB*
Q3 *500 V*
Q4 *(c)*
Q5 *See Section 8.5*
Q6 *p-type*
Q7 *False*
Q8 *Higher current gain; faster switching on and off*
Q9 *85*
Q10 *high*

Chapter 9

Q1 *2*
Q2 *only one input is 1*
Q3 *See Figure 9.3*
Q4 *(b)*
Q5 *See Section 9.4*
Q6 *See Section 9.3*
Q7 *Output is 1 when all three inputs are 1 as the table shows*

A	B	C	S
0	0	0	0
1	0	0	0
0	1	0	0
1	1	0	0
1	0	1	0
0	1	1	0
1	1	1	1

Q8

A	B	$\overline{A}$	$\overline{B}$	P	Q	S
0	0	1	1	1	0	0
1	0	0	1	1	1	1
0	1	1	0	0	1	0
1	1	0	0	0	1	0

Q9 *See Section 9.5*

Q10 (a)

A	B	A + B	A . (A + B)
0	0	0	0
0	1	1	0
1	0	1	1
1	1	1	1

(b)

A	$\overline{A}$	A + $\overline{A}$
0	1	1
1	0	1
1	1	1

(c)

A	B	C	(A + B) . (A + C)	A + (B . C)
0	0	0	0	0
1	0	0	1	1
0	1	0	0	0
1	1	0	1	1
0	0	1	0	0
1	0	1	1	1
0	1	1	1	1
1	1	1	1	1

Chapter 10

Q1 *(b)*
Q2 *See Section 10.2*
Q3 *See Section 10.2*
Q4 *4 kHz*
Q5 *(a) 9999, i.e. 1001 1001 1001 1001 (b) 0000 0011 0110 1001*
Q6 *See Section 10.3*
Q7 *0110 0100 0000 0000 0000*
Q8 *See Section 10.2*
Q9 *See Section 10.3*
Q10 *When the output ABCD reads 1010, i.e. decimal 10, the AND gate produces a logic 1 which resets the counter.*

Chapter 11

Q1 *For inspiration, see Section 11.1*
Q2 *See Section 11.4*
Q3 *For example, using seven-segment displays, liquid crystal displays, and other optoelectronics devices, to display numbers, letters and some punctuation marks and mathematical symbols.*
Q4 *The suppression of a '0' when displaying, say, '2.34' on a four-digit display, rather than '02.34'*
Q5 *E.g. an LCD draws less power than an LED*
Q6 *1 kΩ*
Q7 *See Section 11.2*
Q8 *In terms of a seven-segment display it means switching digits on one at a time in rapid succession so as to deceive the eye that they are all on at the same time.*
Q9 *E.g. advertising banners at football matches...*
Q10 *Power supply 'on' indicator on an electric kettle; a flashing LED on a security alarm system; a torch that can be found in the dark*

Chapter 12

Q1 *Central processor unit; hard disk storage; RAM; input device, e.g, a keyboard*

Q2 *See Section 11.1 and 11.2*

Q3 *To store video clips and photos*

Q4 *A type of computer memory that uses integrated circuits to store data. It is a non-volatile memory so that information stored in it is retained even when the device is switched off. Flash memory cards are used in digital cameras, mobile phones, memory sticks and personal MP3 players.*

Q5 *See Section 12.5*

Q6 *RAM (random-access memory) An integrated circuit that is used for the temporary storage of computer programs. ROM (read-only memory) An integrated circuit that is used for holding data permanently, e.g. for storing the language and graphics symbols used by a computer.*

Q7 *See Section 12.2*

Q8 *A photomask is a transparent glass plate used in the manufacture of integrated circuits on a silicon chip. It contains microscopically small opaque 'spots' that have been produced by the photographic reduction of a much larger pattern. A photoresist is a light-sensitive material that is spread over the surface of a silicon wafer from which silicon chips are made, and whose solubility in various chemicals is altered by exposure to ultraviolet light. A photoresist is used with a photomask so that holes can be etched at selected points in the surface of the silicon.*

Q9 *See Section 12.9*

Q10 *See Section 12.7*

Chapter 13

Q1 *Closed-loop control is a method of controlling the output of a system by feeding its input with an error voltage to reduce the*

difference between the desired and actual outputs. Open-loop
control sets the desired output but as no feedback to monitor
the output and provide a corrective signal to stabilize the output.

Q2 *See Section 13.2*

Q3 *See Section 13.3*

Q4 *See Section 13.4.*

Q5 *See Section 13.4*

Q6 *The open-loop voltage gain of an op amp is the intrinsically high voltage gain unmodified by external components.*

Q7 *(a) 1.5 V; (b) 5 kΩ; (c) 50°C; (d) 1.5 kΩ*

Q8 *By connecting a resistor of value, say, 1 MΩ between the output and the non-inverting input to provide positive feedback*

Q9 *Voltage at X, 4.5 V; at Y, 3 V; at Z, 1.5 V (b) 6 kΩ*

Q10 *See Section 13.7*

Chapter 14

Q1 *E.g. a watch with an analogue display but is fundamentally digital in the way it works*

Q2 *See Section 14.2*

Q3 *See Section 14.2*

Q4 *See Section 14.3*

Q5 *They have an infinitely large open-loop voltage gain. They draw no current whatsoever from the source of the signals at either of their two inputs.*

Q6 *See Section 14.3*

Q7 *(a) This is a non-inverting voltage amplifier; (b) 11; (c) 4.4 V*

Q8 *(a) This is an inverting voltage amplifier; (b) 20; (c) –0.225 V*

Q9 *See Section 14.3*

Q10 *Many environmental properties such as temperature, pressure and wind speed are analogue in that they change continuously. But computers operate on digital signals which have two values, 0 and 1. Hence an analogue-to-digital converter is needed to input data from these external sources into data that a computer can recognise.*

Chapter 15

Q1 *See Section 15.1*

Q2 *gamma rays; X-rays; ultraviolet radiation; infrared radiation; medium frequency radio waves*

Q3 *1/E; 2/C; 3/D; 4/A; 5/B*

Q4 *1.5 m*

Q5 *(b)*

Q6 *(c)*

Q7 *(A) 100; 5 kHz*

Q8 *See Section 15.9*

Q9 *See Section 15.10*

Q10 *100 μs*

Taking it Further

Websites

Portal for the electronics journal, *Everyday Practical Electronics (EPE)*. EPE is a hobbyist magazine for electronics and computer enthusiasts, with construction projects for beginners and experts.

http://www.epemag3.com/

Find out how almost anything works from elements of the human body to space age technology.

http://www.howstuffworks.com

Theory and practice of electronic circuit design and build, organized for beginners and intermediate level enthusiasts.

http://www.doctronics.co.uk/design.htm

Basic theory and practical ideas for the beginner.

http://www.hut.fi/Misc/Electronics/

Intermediate to advanced collection of circuit designs. Includes links to other electronic resources on the web.

http://www.epanorama.net/index.php

Components and circuit designs for schools and industry from Farnell Electronics Ltd.

http://www.farnell.com/

Source of components, electronic kits and books from a major electronics supplier, Maplin.

http://www.maplin.co.uk/

Information about projects, components and books from a major supplier to education, Rapid Electronics Ltd.

http://www.rapideducation.co.uk/products_newproducts.htm

Details about the journal Electronics Education for schools and colleges, published by the Institution for Electrical Engineers, and careers in engineering.

http://www.iee.org/Publish/Adverts/ee.cfm

Keep up to date with the latest news about science and technology from the publishers of New Scientist.

http://www.newscientist.com/

For further information about peripheral interface controllers for educational projects, see the following:

http://www.rev-ed.co.uk/picaxe/

Glossary

A The symbol for the unit of electrical current, the ampere.

ADC (analogue-to-digital converter) A device that changes an analogue signal into a digital signal. Thus, an ADC is used when a computer receives an analogue temperature reading from a thermocouple temperature sensor.

aerial An arrangement of electrical conductors, usually placed in an elevated position (e.g. on or between masts), that transmits and/or receives radio signals.

AF (audio frequency) Any regularly repeating signals that are in the frequency range of human hearing, i.e. from about 20 Hz to 20 kHz.

alphanumerics The use of seven-segment displays, liquid crystal displays, and other optoelectronics devices, to display numbers, letters and some punctuation marks and mathematical symbols.

ammeter An instrument for measuring the strength of an electric current in units of amperes, milliamperes or microamperes.

ampere A flow of electric charge in a circuit equal to 1 coulomb per second.

amplifier A device that increases the voltage, current or power of a signal.

amplitude The strength of a signal measured by its maximum value.

AM (amplitude modulation) A method of sending a message on a radio or light wave by varying the amplitude of the wave in response to the frequency of the message.

analogue Having characteristics that respond to or produce a continuous range of values rather than specific values, e.g. amplitude modulated radio waves.

AND gate A building block in digital logic circuits that makes the following logical decision: it only produces an output of logic 1 when all of its inputs are at logic 1.

anode The terminal of a device (e.g. a diode) from which, electrons flow.

ALU (arithmetic and logic unit) That section of a microprocessor that makes use of logic gates to add and subtract numbers.

AI (artificial intelligence) The study of machines that perform tasks that if done by humans could be considered intelligent. Simple example of artificial intelligence is the ability of a computer to recognize human speech.

aspect ratio A way of describing the width and height of a movie picture. Most TV programmes are broadcast with a 4 : 3 aspect ratio that fits the majority of TV screens in the home. Since this does not match the more visually acceptable 'big screen' aspect ratio, wide screen TVs with 14 : 9 and 16 : 9 aspect ratios are increasingly common.

astable A circuit designed to produce a continuous signal that has a rectangular or square waveform. An astable has numerous applications, e.g. as 'clocks' in digital circuits and in alarm and monitoring systems.

atom the smallest particle of a chemical element that can exist alone or in combination with other atoms.

342

B The symbol for a byte of binary digits comprising 8 bits.

. . . .

bandwidth The range of frequencies contained in a signal, or to which a device responds. Thus a device that responds to the frequency range 0.5 kHz to 4 kHz has a bandwidth of 3.5 kHz.

. . . .

bargraph display An instrument readout that indicates the value of something by the length of a row of glowing light-emitting diodes, or the active elements of a liquid crystal display.

. . . .

base One of the three terminals of a bipolar transistor into or out of which a small base current flows to control a much larger collector current.

. . . .

BCD (binary-coded decimal) A system of representing multi-digit decimal numbers where each digit is represented separately by a four-bit binary number whose values range from 000–101.

. . . .

binary number A number comprising the binary digits, 0 and 1 and having a base of 2.

. . . .

bipolar transistor A transistor that depends for its operation on both n-type and p-type semiconductors, i.e. its function depends on both electrons and holes.

. . . .

bistable (also flip-flop) A circuit that has two outputs that can act as memories for data fed into its input. Flip-flops are used in many types of electronic counter and computer memory.

. . . .

bit (binary digit) Either of the two numbers 0 and 1 that are the basic units of data in computers and other digital systems. A group of bits is known as a word, and a group of 8 bits is known as a byte.

Boolean algebra A shorthand way of dealing with logical statements that are true or false. A 19th century mathematician, George Boole, developed Boolean algebra, and it has since become a useful way of analysing and predicting the behaviour of two-state digital circuits.

boule (also ingot) A very pure crystal of silicon or other semiconductor from which thin slices (wafers) are cut at the start of the complex process of making silicon chips.

breadboard a circuit assembly system enabling components to be connected together temporarily without the need for soldering.

bus The path taken by data and addresses on a computer's motherboard. It connects the processor with different parts of the computer such as the hard disk drive and memory.

byte A group of (usually eight) binary digits, e.g. the byte 10100101, that is handled as a single unit of data in computers and other data handling systems. Desktop and laptop computers typically store gigabytes (GB) of data.

C The symbol for electrical capacitance.

capacitance (C) The ability of a capacitor to store electric charge.

capacitor A component that stores electrical energy by building up electrical charge in a pair of insulated plates.

carrier wave (CW) A relatively high frequency radio or light wave that carries a message from a transmitter to a receiver.

cathode The terminal of a device (e.g. a diode) towards which electrons flow in a circuit.

CCD (charge-coupled device) An integrated circuit comprising an array of minute semiconductor memory cells on a silicon chip that converts light received through a lens into a series of electrical charges that are directly related to the intensity of any given picture element (pixel). CCDs are widely used in security cameras, digital cameras and in telescopes for detecting faint sources such as galaxies.

CD (compact disc) A popular optical disc designed for recording music. A single CD can store 74 minutes of music as a stereo signal.

channel The conducting path between the drain and source terminals of a field-effect transistor.

charge A basic property of matter that occurs in discrete units, usually equal to the charge on an electron, and that can be of positive or negative polarity.

clock Any circuit that produces a regular series of on/off pulses (0s and 1s) that are used to synchronise the flow of binary data in digital processing machines such as computers.

closed-loop control A method of controlling the output of a system by feeding its input with an error voltage to reduce the difference between the desired and actual outputs.

CMOS (complementary metal oxide semiconductor) A switching circuit based on the combination of n-channel and p-channel field-effect transistors. A wide range of digital devices, e.g. digital watches and microprocessors, are based on CMOS devices in integrated circuit form. These chips have the advantage of low power consumption and they can operate from a wide range of supply voltages.

code A set of symbols, e.g. Morse code, or of conventions, e.g. the ASCII code, that represents information in a suitable form for transmission from one place to another.

collector One of the three terminals of a bipolar transistor from which the output is usually taken.

combinational logic A digital circuit, e.g. a NAND gate, that produces an output based on the combination of 0s and 1s presented to its inputs.

comparator An electronic device, e.g. one based on an operational amplifier, that produces an output in response to the difference between the voltages of two inputs.

computer A programmable device used for storing, retrieving and processing data in digital form.

coulomb The unit for measuring electrical charge. A current of 1 ampere is defined as electric charge flowing at the rate of 1 coulomb per second.

counter Any device, e.g. a decade counter, made from flip-flops and used for counting binary data entering its input.

CPU (central processor unit) The principal operating and controlling part of a computer, also known as its microprocessor.

CRO (cathode-ray oscilloscope) A test and measurement instrument for showing the patterns of electrical waveforms and for measuring their frequency and other characteristics.

CRT (cathode ray tube) Once the part of most desktop computer systems and televisions that creates the picture on its screen. They are being replaced by liquid crystal displays and LED displays in many applications.

cycle A complete sequence of a wave pattern that is repeated at regular intervals.

DAB (digital audio broadcasting) A radio system launched in 1995 providing high quality sound and interference-free reception. DAB receivers remain locked onto the broadcast signal when moving around and so are ideal for car radios. DAB can transmit text and graphics along with audio broadcasts.

DAC (digital-to-analogue converter) A device for converting a digital signal into an analogue signal of equivalent value. DACs are found in digital audio devices such as CD, DVD and MiniDisk players for converting digital data to sound.

Darlington pair An arrangement of two bipolar transistors connected in series so that their combined current gain equals (theoretically) the product of their individual current gains. The Darlington pair is widely used in switching and computer interfacing circuits.

decade counter A binary counter that counts up to a maximum of nine before resetting to zero.

decibel (dB) A unit used for comparing the strengths of two signals, such as the intensity of sound and the power gain of an amplifier. The decibel is defined logarithmically so that a doubling of the signal strength is reckoned as an increase of 3 dB, and a halving a decrease of –3 dB.

decoder A device that converts coded information, e.g. the binary code, into a more readily understood code such as decimal.

depletion layer The region across a reverse-biased p–n junction that contains few free electrons and holes and that is responsible for the rectifying properties of a diode.

detector A device in the demodulator stage of a radio receiver that recovers the original signal from the carrier wave.

dielectric The insulating layer of, say, mica, glass or polyester, between the conducting plates of a capacitor.

diffusion The random movement of electrons and/or holes from a region of high to low concentration of these charge carriers.

digital Having characteristics that respond to or produce a numbers especially the two digits 0 and 1, e.g. a digital camera.

digital TV The successor to analogue TV broadcast over cable, satellite and terrestrial systems. Digital transmissions take up less bandwidth than their analogue counterparts so broadcasters can offer more digital channels.

d.i.l. (dual in line)

doping The process of introducing minute amounts of material, the dopant, into silicon to produce n-type or p-type semiconductors in the making of transistors, integrated circuits and other semiconductor devices.

drain One of the three terminals of a field-effect transistor.

driver Any device, e.g. a Darlington pair, that provides sufficient power to operate a load, e.g. a relay when it is called a relay driver.

DVD (digital versatile disc) An extremely versatile 12 cm disc that is used in everything from DVD video players to DVD-ROM drives and recordable DVD+RW machines. DVD is able to store more information than an identically sized CD since the data pits are smaller and more densely packed and are 'read' by a more finely focused laser beam. DVDs use MPEG-2 compression technology to store 4.7 GB of data although for some movies single-sided dual-layer discs are used with a capacity of 8.5 GB.

electrode A metal connector used to make electrical contact with a circuit.

electrolytic capacitor A capacitor that is made from two metal plates separated by a very thin layer of aluminium oxide. Electrolytic capacitors offer a high capacitance in a small volume, but they are polarized and need connecting the right way round in a circuit.

electromagnetic spectrum The family of radiations that all travel at the speed of light through a vacuum, and extend from very short wavelength gamma rays to very long wavelength radio waves, and include light, infrared and ultraviolet radiation.

e.m.f. (electro-motive force) The electrical force generated by a cell or battery that makes electrons move through a circuit connected across the terminals of the battery.

electron A small negatively charged particle that is one of the basic building blocks of all substances and forms a 'cloud' round the nucleus of an atom.

electronics The study and application of the behaviour and effects of electrons in transistors, integrated circuits and other devices.

emitter One of the three terminals of a bipolar transistor that is usually connected to both the input and output circuits.

encoder Any device that converts information into a form suitable for transmission by electronic means.

epitaxial layer One of a number of thin layers of semiconductor that is formed on a layer of silicon in the process of making an integrated circuit by masking and etching.

etching The process of removing silicon dioxide from minute areas of a silicon chip during the several stages involved in the making of an integrated circuit. Following etching, the underlying layer of silicon is doped with an impurity in order to change its electrical properties to an n-type or p-type semiconductor.

exclusive-OR gate A building block in digital logic circuits that makes the following logical decision: it produces an output of logic 1 when any one, but not all, of its inputs are at logic 1.

farad (F) The unit of electrical capacitance and equal to the charge stored in coulombs in a capacitor when the potential difference across its terminals is 1V.

feedback The sending back to the input of part of the output of a system in order to improve the performance of the system. Feedback can be either positive or negative.

fibre optics The use of hair-thin transparent glass fibres to transmit information on a light beam that passes through the fibre by repeated internal reflections from the walls of the fibre.

FET (field-effect transistor) A transistor depending for its operation either on n-type or on p-type semiconductor material. The FET is a unipolar device depending either on the flow of electrons, as in an n-type FET, or holes, as in a p-type FET.

filter A device for controlling the range of frequencies passing through a circuit. For example, low-pass and high-pass filters are used in a crossover circuit in a hi-fi system.

flash memory A type of computer memory using integrated circuits to store data. It is a non-volatile memory so that information stored in it is retained even when the device is switched off. Flash memory cards are used in digital cameras, mobile phones, memory sticks and personal MP3 players.

flip-flop (see bistable)

forward bias A voltage applied across a pn junction that causes electrons and holes to flow across the junction.

frequency The number of times per second that a regular process repeats itself. Frequency is measured in hertz (Hz) or cycles per second. The frequency of the mains power supply is 50 Hz.

FM (frequency modulation) A method of sending information by varying the frequency of a radio in response to the amplitude of the message being sent. For high quality radio broadcasts, FM is preferable to AM since it is affected less by interference from electrical machinery and lightning.

full-wave rectifier A semiconductor device based on four diodes that produces direct current from alternating current by reversing the flow of current in one half cycle of the alternating current.

fuse A device that acts as the 'weak link' in a circuit that it protects from excessive current flow.

gain The increase in the power, voltage or current of a signal as it passes through an amplifier.

gallium arsenide (GaAs) A crystalline material that, like silicon, is used to make diodes, transistors and integrated circuits. GaAs conducts electricity about seven times more easily than silicon, and it is favoured for computer memory devices used in guided missiles and other fast-acting weapons.

gamma rays Electromagnetic radiation having a much shorter wavelength than light. Gamma rays are generated by radioactive substances and are present in cosmic rays.

gate One of the three terminals of a field-effect transistor that controls current flow between its drain and source terminals.

germanium A non-metallic element of atomic number 32 and the first material to be used as the basis of transistors and diodes.

gigabyte (GB) A quantity of computer data equal to one thousand million (American billion) bytes.

gigahertz (GHz) A frequency equal to one thousand million hertz (10^9 Hz).

the unit of H The symbol for, the electrical inductance henry.

half-wave rectifier A diode, or circuit, based on one or more diodes, that produces a direct current from alternating current by removing one half of the AC waveform.

heat sink A relatively large piece of metal that is placed in contact with a transistor or other component to help dissipate unwanted heat generated within the component.

HDTV (high definition television) A system for displaying more picture information by using progressive scan techniques that deliver either 720 or 1080 lines rather than the European 625-line standard. All HDTV broadcasts are in widescreen format.

henry (H) The unit of electrical inductance defined as the potential difference generated across the terminals of an inductor when the current through it is changing at the rate of one ampere per second.

hertz (Hz) The unit of frequency equal to the number of complete cycles per second of an alternating waveform.

hole A vacancy in the crystal structure of a semiconductor that is able to attract an electron. A p-type semiconductor contains an excess of holes acting as mobile charge carriers which move through the semiconductor under the action of an electric field.

impurity An element such as boron that is added to silicon to produce a semiconductor with desirable electrical qualities.

inductance (L) The property of a circuit, especially a coil of wire, that makes it generate a voltage when a current, either in the circuit itself or in a nearby circuit, changes.

inductor An electrical component, usually in the form of a coil of wire, that is designed to resist changes to the flow of current through it. Thus inductors are used as 'chokes' to reduce the possibly damaging effects of sudden surges of current, and in tuned circuits.

infrared Radiation having wavelengths between the visible red and microwave regions of the electromagnetic spectrum, i.e. wavelengths between about 700 nanometres (700×10^{-9} m) and 1 mm. Though invisible to the naked eye, infrared is widely used in electronics, especially in remote control and security systems.

input The point at which information enters a device.

insulator A material, e.g. glass, that does not allow electricity to pass through it.

IC (integrated circuit) An often very complex electronic circuit that has resistors, transistors, capacitors and other components formed on a single silicon chip.

interface A circuit or device, e.g. an ADC, that enables a computer to transfer data to and/or from its surroundings or between computers.

ion An atom or group of atoms that has gained or lost one or more electrons and that therefore carries a net positive charge.

junction A region of contact between two dissimilar metals (as in a thermocouple) or two dissimilar semiconductors (as in a diode) that has useful electrical properties.

k The symbol used for the prefix 'kilo' meaning one thousand times.

kilobit (kb) One thousand bits, i.e. 0s and 1s, of data.

kilobyte (kB) One thousand bytes of data.

kilohertz (kHz) A frequency equal to 1000 Hz.

laser A device that produces an intense and narrow beam of light of almost one particular wavelength. The light from lasers is used in optical communications systems, CD and DVD players.

LCD (liquid crystal display) A display used in computer monitors, notebook PCs and TVs that is typically smaller, lighter and less power hungry than their cathode ray tube equivalents. An LCD operates by controlling the reflected light from special liquid crystals, rather than by emitting light as in the light-emitting diode. Most LCDs now use active matrix thin-film technology that marries each pixel with its own transistor.

LDR (light-dependent resistor) A semiconductor device that has a resistance decreasing sharply with increasing light intensity. The LDR is used in light control and measurement systems, e.g. automatic street lights and cameras.

LSB (least significant bit) The right-most digit in a binary word, e.g. 1 is the LSB in the word 0101.

LED (light-emitting diode) A small semiconductor diode that emits light when current passes between its anode and cathode terminals. Red, green, yellow and blue LEDs are used in all types of display systems.

load The general name for a device, e.g. an electric motor, under the control of an electronic circuit, that absorbs electrical energy to produce mechanical or some other form of energy.

logic diagram A circuit diagram showing how logic gates and other digital devices are connected together to produce a working circuit or system.

logic gate A digital device, e.g. an AND gate, that produces an output of logic 1 or 0 depending on the combination of 1s and 0s at its inputs.

loudspeaker A transducer device used to convert electrical energy into sound energy. It usually comprises a coil of wire, attached to a paper cone, and which is located in a strong magnetic field. The coil and cone move when current flows through the coil.

m The symbol for the prefix 'milli' meaning one thousandth of.

M The symbol for the prefix 'mega' meaning one million times, e.g. megabytes (20 MB).

majority charge carrier The more abundant of the two charge carriers in a semiconductor. The majority charge carriers in n-type material are electrons.

medium waves Radio waves having wavelengths in the range about 200 to 700 m, i.e. frequencies in the range 1.5 MHz to 4.5 MHz.

MHz (megahertz) A frequency equal to one million (10^6) Hz. The performance of a computer is often stated in megahertz as a measure of the speed of the computer's clock. However, it is not the only factor determining the computer's performance.

megapixel Equal to one million picture elements, or pixels. This is a measurement applied to the image resolution of a CCD device such as digital cameras and scanners.

memory That part of a computer system used for storing data until it is needed. A microprocessor in a computer can locate and read each item of data located in memory.

memory stick A very compact form of data storage for portable devices such as notebook computers, digital cameras and camcorders. The original Memory Stick designed by Sony was about the size of a highlighter pen or a pack of chewing gum.

microcontroller: A programmable computer packaged in a single IC chip used for electronic control applications.

microelectronics The production and use of complex circuits integrated on silicon chips.

microfarad (µF) A unit of electrical capacitance equal to one millionth of a farad.

micron (micrometre) A distance equal to one millionth of a metre (µm). The micron is used for measuring the size and separation of components on silicon chips.

microphone A transducer that converts sound waves into electrical signals usually for subsequent amplification.

microprocessor A complex integrated circuit manufactured on a single silicon chip and also known as a central processor unit. It is the 'heart' of a computer and can be programmed to perform a wide range of functions. A microprocessor is used in washing machines, cars, cookers, games and many other products.

microsecond (μs) A time interval equal to one millionth (10^{-6}) of a second.

microswitch A small mechanically operated switch that usually needs only a small force to operate it.

microwaves Radio waves having wavelengths less than about 300 mm and used for straight line communications.

millimetre (mm) A distance equal to one thousandth (10^{-3}) of a metre.

millisecond (ms) A time equal to one thousandth (10^{-3}) of a second.

milliwatt (mW) A power equal to one thousandth (10^{-3}) of a watt.

MiniDisk A digital alternative to the analogue compact cassette launched by Sony in 1992. The disc comprises a 64 mm diameter magneto-optical disc capable of storing 80 minutes of music in a standard play mode. The advantage over compact discs is that the user can name, edit, split and combine individual tracks and put them in any order without erasing the disc and starting again.

minority charge carrier The less abundant of the two charge carriers present in a semiconductor. The minority charge carriers in n-type material are holes.

modulator A circuit that puts a message on some form of carrier wave, e.g. light waves, for transmission in a communications system.

Monostable A circuit that produces a time delay when it is triggered, and then reverts back to its original, stable, state in readiness for a subsequent trigger pulse.

MSB (most significant bit) The left-most digit in a binary word, e.g. 0 is the MSB in the binary word 0111.

MP3 A digital codec (compression/decompression) based on MPEG-1 enabling audio files to be compressed at different levels of quality. MP3 enables users to trade off sound quality against the amount of storage space available on a PC or personal MP3 player.

MPEG (Moving Picture Experts Group) The group that developed the technology of MPEG compression/decompression (codec) for audio and visual data. MPEG-1 was launched in the early 1990s but its successor MPEG-2 has been adopted for use in DVD and digital TV systems. The variant MP3 is used for Internet audio, while MPEG-4 is an Internet standard for the streaming and standalone playback of video clips on computers.

multimeter An analogue or digital instrument for measuring current, potential difference and resistance and used for testing and fault-finding in designing and testing of electronic circuits.

multiplexing A method of making a single communications channel carry several independent messages. In terms of a seven-segment display it means switching digits on one at a time rapidly so as to deceive the eye that they are all on at the same time.

n The symbol for the prefix 'nano' (nm) meaning one thousand millionth of.

NAND gate A building block in digital logic circuits that makes the following logical decision: it produces an output of logic 1 when one or more of its inputs are at logic 0, and an output of logic 0 only when all its inputs are at logic 1.

nanofarad (nF) A unit of electrical capacitance equal to one thousand millionth (10^{-9}) of a farad.

nanosecond (ns) A time interval equal to one thousand millionth (10^{-9}) of a second.

negative feedback The feeding back to the input of a system part of its output signal so as to reduce the effect of the input. Negative feedback is widely used in amplifiers and control systems to improve their stability.

neutron A particle in the nucleus of an atom that has no electrical charge and a mass roughly equal to that of the proton.

NOR gate A building block in digital logic circuits that makes the following logical decision: it produces an output of logic 1 only when all its inputs are at logic 0, and an output of logic 0 when one or more of its inputs are at logic 1.

NOT gate A building block in digital logic circuits that makes the following logical decision: it produces an output of logic 1 when its single input is at logic 0, and vice versa.

npn transistor A semiconductor device that is made from both p-type and n-type semiconductors and that is used in switching and amplifying circuits. An npn transistor has three terminals: emitter, collector and base. A small current flowing into its base terminal controls a larger current flowing between the emitter and collector terminals.

n-type semiconductor A semi conductor through which current flows mainly as electrons.

nucleus The central and relatively small part of an atom that is made up of protons and neutrons.

ohm (Ω) The unit of electrical resistance.

Ohm's law The potential difference across the ends of metallic conductor is proportional to the current flowing through it if its temperature and other physical factors remain unchanged.

open-loop control A means of setting the output of a system without providing negative feedback to ensure the output remains at the set level.

operational amplifier (op amp) A very-high-gain amplifier producing an output voltage that is proportional to the difference between its two input voltages. Operational amplifiers are widely used in instrumentation and control systems.

optical communications The use of long thin optical fibres for sending messages using pulses of laser light.

optical fibre A thin glass or plastic thread through which light travels without escaping from its surface. They commonly feature in decorative lamps and festive decorations but have their primary use in optical communications systems.

optoelectronics A branch of electronics dealing with the interaction between light and electricity.

OR gate A building block in digital logic circuits that makes the following logical decision: it produces an output of logic 1 when one or more of its inputs are at logic 1, and an output of logic 0 when all its inputs are at logic 0.

oscillator A circuit or device providing a sinusoidal or square wave signal.

PAM (pulse amplitude modulation) A type of signal modulation in which the amplitude of individual, regularly spaced pulses in a pulse train is varied in accordance with the amplitude of the modulating signal.

p The symbol for the prefix 'pico' meaning one million millionth of (10^{-12}).

parallel circuit A circuit in which components, e.g. resistors, are connected side-by-side so that current is shared by the components.

PCM (pulse code modulation) A method of sampling an analogue waveform and converting the sample into a digital signal. For example, analogue waveforms to be stored on a CD are converted into a series of 0s and 1s at 16 bit/ 44.1 kHz resolution. The sampling rate determines the ability of the CD to represent music in its original analogue form.

period The time taken for a wave motion, e.g. a radio wave, to make one complete oscillation.

photodiode A light-sensitive diode that has two terminals, an anode and a cathode, and that responds rapidly to changes of light.

photomask A transparent glass plate used in the manufacture of integrated circuits on a silicon chip. It contains microscopically small opaque 'spots' that have been produced by the photographic reduction of a much larger pattern.

photon The smallest 'packet', or quantum, of light or other types of electromagnetic energy.

photoresist A light-sensitive material that is spread over the surface of a silicon wafer from which silicon chips are made, and whose solubility in various chemicals is altered by exposure to ultraviolet light. A photoresist is used with a photomask so that holes can be etched at selected points in the surface of the silicon.

PIC chip A brand of microcontroller manufactured by Microchip Technology Inc. (PIC: peripheral interface controller)

picofarad (pF) An electrical capacitance equal to one million millionth of a farad (10^{-12} F).

piezoelectricity The electricity that certain crystals (e.g. quartz) produce when they are squeezed. Conversely, if a potential difference is applied across a piezoelectric crystal, it alters shape slightly. The piezoelectric effect is put to good use in ultrasonic transducers, crystal microphones and gas lighters.

pnp transistor A semiconductor device that is made from both p-type and n-type semiconductors and that is used in switching and amplifying circuits. A pnp transistor has three terminals, emitter, collector and base. The current flowing between the emitter and collector terminals is controlled by a small current flowing out of the base terminal.

port A place on a computer to which peripherals can be connected to provide two-way communication between the computer and the outside world.

positive feedback The feeding back to the input of a system of a part of its output signal so as to increase the effect of its input. Positive feedback is used in an astable.

potential divider Two or more resistors connected in series through which current flows to produce potential differences dependent on the resistor values.

potentiometer An electrical component having three terminals that provides an adjustable potential difference.

proton A particle that makes up the nucleus of a hydrogen atom, that coexists with neutrons in the nuclei of all other atoms, and has a positive charge equal in value to the negative charge on an electron.

p-type semiconductor A semiconductor through which current flows mainly as holes.

pulse A short-lived variation of voltage or current in a circuit, or of an electromagnetic wave such as laser light.

R The symbol for electrical resistance, the ohm.

radar (RAdio Detection And Range-finding) An electronic system for locating the position of distant objects by recording the echo of high frequency radio waves bounced off the object.

radio Communication at a distance by means of electromagnetic waves having frequencies in the range about 15 kHz to 30 GHz.

RAM (random-access memory) An integrated circuit that is used for the temporary storage of computer programs and data.

rectifier A semiconductor diode that makes use of the one-way conducting properties of a p–n junction to convert a.c. to d.c.

relay A magnetically operated switch that enables a small current to control a much larger current in a separate circuit.

resistor A component that offers resistance to electrical current and across which there is a potential difference.

reverse bias A voltage applied across a p–n junction (e.g. a diode) that prevents the flow of electrons across the junction.

robot A computer-controlled device that can be programmed to perform repetitive tasks such as paint-spraying, welding, and machining of parts. However, artificial intelligence enables robots to perform decision-making tasks.

ROM (read-only memory) An integrated circuit that is used for holding data permanently, e.g. for storing the language and graphics symbols used by a computer.

s The symbol for a second of time.

Schmitt trigger A snap-action electronic switch using positive feedback that turns off and on at two specific input voltages. The Schmitt trigger is widely used to 'sharpen up' slowly changing waveforms, and to eliminate noise from signals input to circuits such as thermostats and counters.

Schottky A family of logic circuits in integrated circuit packages that is compatible with the older transistor-transistor logic (TTL).

semiconductor A solid material that is a better electrical conductor than an insulator (e.g. polythene), but not such a good conductor as a metal (e.g. silver). Diodes, transistors and integrated circuits are based on n-type and p-type semiconductors.

sensor Any device (e.g. a thermistor) that produces an electrical signal indicating a change in its surroundings (e.g. temperature).

sequential logic A digital circuit that can store information. Sequential logic circuits are based on flip-flops and are the basic building blocks of digital counters and computer memories.

servosystem An electromechanical system that uses sensors to precisely control and monitor the movement of something (e.g. the read/write head of a CD player).

short waves Radio waves that have wavelengths between about 2.5 MHz and 15 MHz, and that are mainly used for amateur and long-range communications.

silicon An abundant non-metallic element that has largely replaced germanium for making diodes and transistors. Silicon is doped with small amounts of impurities such as boron and phosphorus to make n-type and p-type semiconductors.

silicon chip A small piece of silicon about the size of this letter 'o' containing a complex miniaturised circuit (called an integrated circuit) formed by photographic and chemical processes.

SMD (surface mounted device) A component, e.g. an integrated circuit, that is designed for soldering in place on one side of a printed circuit board.

stepper motor An electric motor with a shaft that rotates one step at a time taking, for example, 48 steps to complete one revolution. Stepper (or stepping) motors are used for the precise positioning of robot arms and printer paper, for example, under computer control.

strain gauge A sensor attached to an object to detect how much it lengthens or shortens when it is loaded. The small change in electrical resistance of the strain gauge is a measure of the distortion produced.

telecommunications The use of electronic and other equipment to send information through wires, the air and interplanetary space.

thermistor A device made from a mixture of semiconductors such that its resistance changes with temperature. Thermistors are used in control systems such as thermostats and for temperature measurement.

thermocouple A device made from a pair of dissimilar metals (e.g. copper and iron) that produces a voltage varying with temperature. Thermocouples are ideal as sensors in electronic thermometers since the junction between the metals can be made very small.

transducer A device (e.g. a thermocouple, a solar cell or a loudspeaker) that changes one form of energy into another. Transducers are widely used in electronic sensing and control systems.

transformer An electromagnetic device for converting alternating current from one voltage to another.

transistor A semiconductor device (e.g. a bipolar transistor) that has three terminals and is used for switching and amplification.

TTL (transistor–transistor logic) A type of digital IC based on bipolar transistors, that provides logic and counting functions and requires a 5 V supply.

truth table A table of 0s and 1s that shows how a digital logic circuit (e.g. a NAND gate) provides a digital output in response to all possible combinations of binary input signals.

tuned circuit A circuit that can be tuned to select particular radio signals.

UHF (ultra-high frequency) Radio waves that have frequencies in the range 500 MHz to 30 GHz and that are used for TV broadcasts.

ultrasonic waves Sound waves inaudible to the human ear that have frequencies above about 20 kHz.

unipolar transistor A transistor that depends for its operation on either n-type or p-type semiconductor materials as in a field-effect transistor.

USB (universal serial bus) A universal connection for computers and associated peripherals such as keyboards, printers and scanners. USB supports data transfer speeds of up to 12 megabits per second (Mbps). It also enables 'plug and play' so that additional software is not needed. USB 2.0 shares USB's capacity to support up to 128 devices from a single port, and also offers transfer of data at up to 480 Mbps.

VHF (very high frequency) Radio waves that have frequencies in the range 30 MHz to 300 MHz and that are used for high radio broadcasts, especially FM.

virtual earth The provision of a connection in a circuit which is assumed to be grounded at 0V but is not actually connected to ground.

volt (V) The unit of electrical potential difference that causes a current of one ampere to flow through a resistance of one ohm. More precisely, it is defined as the energy given to one coulomb of charge so one volt equals one joule per coulomb.

voltage The number of volts at a point in a circuit relative to the circuit's zero potential.

voltmeter An instrument for measuring the electrical potential in volts between two points in a circuit.

W The symbol for the watt, the unit of power.

wafer A thin disc cut from a single crystal of silicon on which hundreds or thousands of integrated circuits are made before being cut up into individual ICs for packaging.

watt (W) The unit of power and equal to the rate of conversion of energy of 1 joule per second.

waveform The shape of an electrical signal, e.g. a sinusoidal waveform.

wavelength The distance between one point on a wave and the next corresponding point (e.g. from crest to crest). Wavelength is related to the frequency and the speed of the wave by the simple equation, speed = wavelength × frequency.

Wheatstone bridge An arrangement of four resistors used for resistance-to-voltage conversion in some types of electronic instrument.

X-rays Penetrating electromagnetic radiation used in industry and in medicine for seeing below the surface of solid materials.

zener diode A semiconductor diode designed to conduct current in the reverse-bias direction at a particular reverse-bias voltage. Zener diodes are widely used to provide stabilized voltages in electronic circuits.

Index

Credits

Front cover: © Photodisc/Getty Images

Back cover: © Jakub Semeniuk/iStockphoto.com, © Royalty-Free/Corbis, © agencyby/iStockphoto.com, © Andy Cook/iStockphoto.com, © Christopher Ewing/iStockphoto.com, © zebicho – Fotolia.com, © Geoffrey Holman/iStockphoto.com, © Photodisc/Getty Images, © James C. Pruitt/iStockphoto.com, © Mohamed Saber – Fotolia.com